Acknowledgements

Writing Group Leaders
Mr J Bradley, Mr R Fitzpatrick, Mr D Gray, Mr C Lambert, Mr M Riches

Editorial Group
Mr R Fitzpatrick, Mr J Goddard, Mr J Godfrey, Mr R Hyde, Dr M Keating, Mr J Tafler

Southern Writing Group
Mr R Fitzpatrick, Ms R Bowditch, Ms C Fry, Mr J Goddard, Mr J Godfrey, Mr R Hyde,
Dr M Keating, Mr J Tafler

Northern Writing Group
Mr M Riches, Mr B Bibby, Ms E Carson, Mr R Carter, Ms B Ditchfield, Mr A Glover

Hertfordshire Writing Group
Mr J Bradley, Ms J M Lane, Mr W Merrick, Mr E Parry

Berkshire Writing Group
Mr D Gray, Mr J Maple, Mr D Williams

Buckinghamshire Writing Group
Mr C Lambert, Ms M Byrne, Mr M Cowell, Ms A Curry, Mr M Dygnas, Mr C Frost,
Mr N Green, Ms M Hannon, Ms E Hunter, Ms A Lathrope, Mr D Lees, Ms A Reynolds,
Ms C Richardson

To   Bill Hall
     Brian Mowl
     John Bausor
     David Wimpenny
     . . . who started the ball rolling with their first 'Patterns' series;

     Kevin Keohane
     Jim Harrison
     . . . who gave us such support when we began;

     Bob Fairbrother
     . . . who not only read the manuscripts with enormous care but showed us how
     best to put right our goofs;

     Marie Allen
     Sally Lyth
     Karen Ryan
     . . . who typed the manuscripts;

     David Lock
     . . . who created the cartoons.

Mike Lyth

Publisher's acknowledgements

We are grateful to the following for permission to reproduce photographs and original artwork:
Airship Industries, 6.10; Randa Bishop/Colorific, 7.49(a); Biophoto Associates, 4.9(b and c), 5.25 and 7.10; British Geological Survey, 2.10, 2.11, 2.13, 6.19 and 6.50(a); British Sidac, 6.23; British Steel, 2.21 (Brian Richards); Camera Press, 2.19 (Steve Benbow/Network); Canadian High Commission, 6.44; J. Allan Cash, 7.3(a); Caterpillar/H. Leverton, 7.3(b); John Cleare/Mountain Camera, 3.19(c); Compix, 3.29(b); James Davis, 6.12; De Beers Diamond Information Centre, 6.38(c) and 6.45(c); Horace Dobbs, 3.19(a); Dunlop, 7.6(a); Everest Double Glazing/Insulation Glazing Association, 3.29(a); *Farmers Weekly*, 3.34(a) (Keith Huggett), (b) (Niall Rankin), (c) (Peter Allen), 3.35(a) (Gordon Craddock), (b) and (c) (Ben Tyrer), and 6.22 (Peter Allen); Martyn Farr, 5.1(a); Vivien Fifield, 1.15 and 7.24; Goodyear, 2.7 and 6.26 (Magna Pictures); Griffin and George, 5.31; Tom Hanley, 6.5; Robert Harding Associates, 7.1 (*Origins*, Richard Leakey, Macdonald); Richard Hyde, 7.6(c), 7.21, 7.30 and 7.32; International General Electric Medical Systems, 6.45(e); Keystone Press, 5.1(b) and 7.23; Clive Lambert, 3.38, 3.40, 7.17, 7.31, 7.33 and 7.35; Frank Lane Agency, 2.1 (US Navy), 3.1(a) (Karl H. Maslowski), (b) (Lynwood M. Chace), (c) (A. Faulkner Taylor), (d) (Jean-Marie Baufle), 3.15 (R. Van Nostrand), 3.16 (Hermann Schunemann), 3.17 (Georg Nystrand), 3.18 (Irene Vandermolen), 3.20 (R.S. Virdee), 3.27 (R.P. Lawrence), 3.28 (Irene Vandermolen) and 7.44 (Leonard Lee Rue); Mansell Collection, 2.4; Mary Glasgow Publications, 1.33 (*Quest 1*, Series 1); National Coal Board, 7.43 (Eagle Alexander); Natural Science Photos, 1.20 (P.H. Ward) and 2.9 (C.A. Walker); *New Scientist*, 7.49(b); Peter Newark's Western Americana, 6.30; Oxford Scientific Films, 3.32 (Cindy Buxton and Annie Price/Survival Anglia) and 4.1 (David Thompson); Pictorial Press, 4.10 and 4.13; Picturepoint, 4.9(a) and 7.20(b); Plessey Research, Caswell, 6.47; Press-Tige Pictures, 6.8, 6.24 and 6.28; Rio Tinto Zinc, 2.22; Ann Ronan Picture Library, 3.44, 7.40 and 7.41; Royal Astronomical Society, 1.12; Royal Commission on Historic Monuments, 2.17 and 2.18; Science Photo Library, 4.21 and 5.6 (Jeremy Burgess, John Innes Institute); Scott Polar Research Institute, 3.19(b); Sefton Photo Library, 6.38(a) and 7.26; Shell UK, 5.18; Harry Smith, 3.36 (R.H.M. Robinson); Space Frontiers, 1.14 and 7.6(d) (NASA); Sport and General, 3.19(d) (Dave Bunce); Sporting Pictures UK, 7.6(b) and 7.34; University of Dundee Electronics Laboratory, 1.34; United Kingdom Atomic Energy Authority, 1.13 and 7.25; Unilever, 1.4, 1.5 and 1.6. The cover photograph of soap bubbles is by David Parker/Science Photo Library.
We are also grateful to the following for permission to redraw copyright material:
BBC, *Exploring Science* Teachers' Notes, 1.16 and 5.28, *Project*, 1.17 and *Science Session*, 1.32; The Diagram Group and *Penguin Book of the Physical World*, 2.2, 6.31 and 7.25; Dover Publications Inc and *The Restless Universe*, Max Born, 1951, 5.9; Lucas Medical, 5.27; Mary Glasgow Publications, *Quest*, 1.11, 1.31 and 2.24; Nuffield Foundation, Revised Nuffield Physics, *Pupils' Text Years 1 and 2*, 1978, 1.18 and 1.23, Nuffield Secondary Science, *Theme 6 Movement*, 1971, 1.22, Nuffield Physics, *Guide to Experiments III*, 1967, 5.5, Revised Nuffield Chemistry, *Teachers' Guide II*, 1978, 6.35, Revised Nuffield Chemistry, *Chemists in the World*, 1979, 5.16.

Contents

7 Energy

Preface

A message to our readers

What is a scientist?

What is a scientist? What do scientists do?
What makes their work different from that of other people?
What do scientists know about?
What great ideas have they developed?
What do they want to find out about next?

The books in the *Exploring Science* series set out to answer these questions –
especially the first two. Mostly the scientists are in the background – though
you will meet one or two in person as you turn the pages. We hope you will
enjoy sharing their great ideas and their special ways of looking at the world.
We hope you will enjoy learning about their special equipment and methods –
and that you will be able to try some of these for yourself. Most of all we hope
that the books will help you to feel what it is like to be a scientist – and how
exciting it can be!

Patterns

As you read the books, you will notice that the word *'pattern'* crops up quite
often. Scientists like to arrange all the things and happenings in the world
into groups with something in common. They like to arrange things to make
patterns. In Chapter 3 you will read about some of the patterns they have
made to do with living things. In Chapter 5 the patterns will be to do with
gases, liquids, and solids, and tiny invisible particles called molecules.
Chapter 1 has a lot to say about the whole idea of patterning. For instance,
it explains why patterning is interesting and useful to scientists.

In your notebook

Every two or three pages through the books you will come across a heading
which says: *'In your notebook'*. Next to it you will find a list of things to do –
an assignment. Usually it will mean writing something, but it might involve
making a drawing, finding a picture to stick in your book, or completing a
table. If you complete all the assignments in your notebook, you will end up
with a valuable collection of all the key ideas in the books. It will be
something to be proud of – if you have kept it neat and tidy!

In some places, we have left gaps for you to fill in. That is partly to make
sure that you have read the pages first! Don't worry if you find the gap-
filling easy – it should be if you have carefully read and understood the
pages.

Checkout

At the end of every chapter you will find the 'Checkout pages'. There are
four parts to these. First, there is a list of all the important words in the

chapter. You should be able to say what each of the words means. You might like to make up a 'scientists' dictionary' in the back of your notebook.

Next, there is a list of the patterns which the chapter has introduced. You might like to copy these into your notebook at the end of each of your chapters.

Then there are ten true–false questions. You can use these to check your understanding of the chapter. If you really can't decide whether a statement is true or false, your science teacher will be glad to discuss it with you.

Finally, there are two 'problems'. They are always connected with the ideas and patterns in the chapter, but usually they give information which you have not seen before. You need to use the *old* patterns to organise *new* information. You may find the problems difficult – interesting scientific problems usually are – but break them up and work at them step by step and you should be successful. Again, if you really are stumped, your teacher will be glad to help.

A last word!

We enjoyed writing the books. We have tried to make them interesting. We hope you enjoy reading them.

The Authors

1 Introduction

1.1 What are scientists like?

A few years ago, a group of nine-year-old children were asked to write a story called 'A Day in the Life of a Scientist'. The children's teacher wanted to find out how they pictured scientists in their minds.

When the stories were written, the children and the teacher had a discussion. The first thing to come out was that all the children had written about men. As far as they were concerned, there were no women scientists! The cartoons show some other things which the nine-year-olds agreed about. . . .

'Scientists . . .
– never get into debt
– do not like going out for a drink
– do not like fast cars
– do not smoke.'

Fig. 1.1

'Scientists . . .
– work long hours
– are not affectionate or sociable
– do not flirt.'

Fig. 1.2

'Scientists . . .
– do not give lots of presents
– do not like expensive restaurants
– are happy to do household chores.'

Fig. 1.3

Do you think the children were right? Do you think scientists are really like this?

Do you know any scientists? What about your science *teachers*? Do they fit in with the nine-year-olds' opinions? Why not ask them?!

Are there any things that are true of *all* scientists – or are they just as different from each other as everyone else?

We can look at a few of today's young scientists . . .

Susan joined the Unilever Research Lab at Port Sunlight in Lancashire

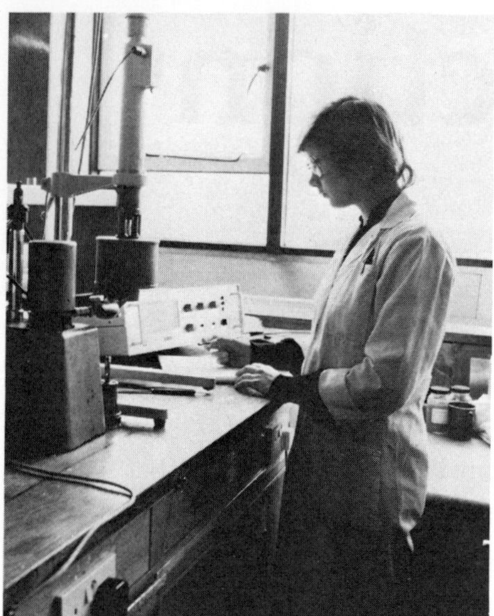

Fig. 1.4 Susan – measuring the viscosity of a fabric conditioner

Fig. 1.5 Jane – showing the progress of young plants grown in her genetics experiment

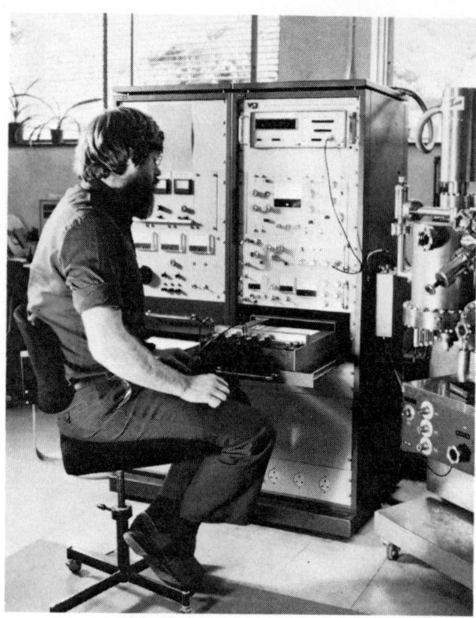

Fig. 1.6 Bob – using an electron spectrometer to investigate a hard surface

straight from school when she was 16. She had seven O-Levels including three in science subjects. She enjoys analysing the results of her work and the computing she becomes involved in.

She says, 'I am sociable and I enjoy going out. Of course some of the scientists I work with do not enjoy doing the same things but we are all different'.

In the picture, Susan is measuring the 'runniness' (called viscosity – see page 9) of a fabric conditioner. Fabric conditioners are used when rinsing clothes to soften them and cut down the static electricity on man-made fibres.

Jane took A-Levels in two science subjects and went to Liverpool University where she took a degree in genetics. (Genetics is the study of the ways that characteristics are passed on from parents to children – see page 77.) Now she works at the Unilever Research Lab near Bedford as part of a team in 'Applied Biosciences'.

She says, 'There are many different kinds of scientists. Some of them *are* very quiet and do not like busy social lives, but others are just the opposite. The people I work with all enjoy doing different things'.

On leaving school, Bob started work as a laboratory technician in a university. He joined Unilever at Port Sunlight in 1974. Whilst there, he has taken a day release course leading to a Higher National Certificate in Chemistry.

He says, 'I do quite a lot of things on my own, like playing classical music on the guitar, but I enjoy playing squash too. You cannot generalise about people just because they do similar work'.

In the picture, Bob is using an instrument called an 'electron spectrometer' to investigate a hard surface. Information from the instrument tells him about the chemical make-up of the surface. The information is used in developing cleaning powders.

In fact, the children were wrong about scientists. Scientists are as different in their likes and dislikes and in their habits and hobbies as any other group of people. However, when it comes to their work, they do have a great deal in common. You will read about that in the rest of the chapter.

1.2 Expert observers

Scientists are experts at looking and seeing – they are expert observers. They spend much of their lives looking carefully at the world and all the things and happenings in it. They make careful notes about their observations.

When scientists see something interesting, they look around for something else to observe in the same place or at the same time. They like to pair up their observations. Double observations are easier to check and less likely to be wrong than single observations. When all scientists agree abqut one of these double observations, it is called a *fact*.

You can see the difference between ordinary observations and scientific observations when a crime has been committed (see Fig. 1.7).

Fig. 1.7

It's easy to make a mistake of observation – especially if you're upset.

It's much less easy to make mistakes about observations of bloodstains, footprints or fingerprints on broken glass. Such 'evidence' is collected for the police by *forensic* scientists. The information which comes back from them will always say something about *comparisons* between *pairs* of observations.

You can do some 'forensic science' for yourself. Read about fingerprints, then try the investigations.

1.3 Fingerprints

No one in the world has fingerprints just like yours. They were formed six months before you were born, and they will be with you, unchanged, for the rest of your life. If you lose some skin in a scrape or burn, the patterns will be there underneath and they will grow back.

About a hundred years ago, the scientist Francis Galton (see page 21) was the first to organise fingerprints into a system for identifying people. By 1911, the taking of 'prints' had helped to solve many crimes, and the police had started to keep records.

The first step in organising a fingerprint file is to pick out the most common patterns and give them names. Simplified versions of these are

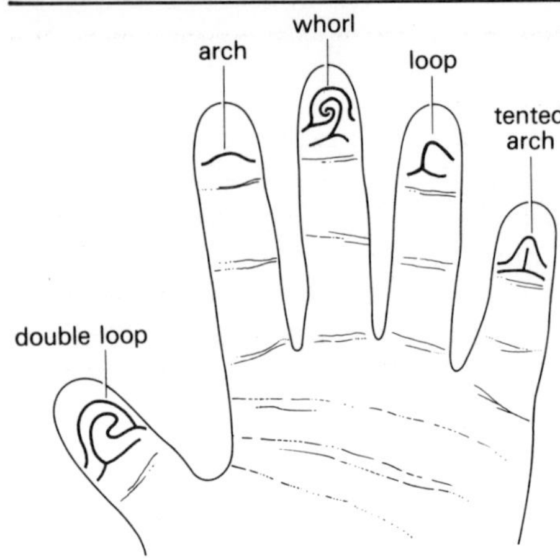

Fig. 1.8 The most common fingerprint patterns

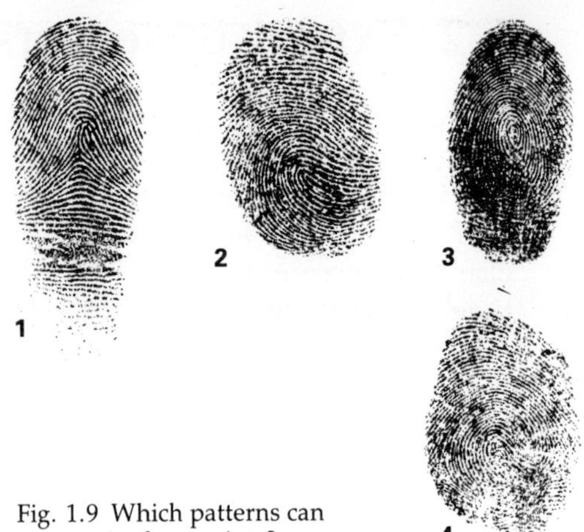

Fig. 1.9 Which patterns can you see in these prints?

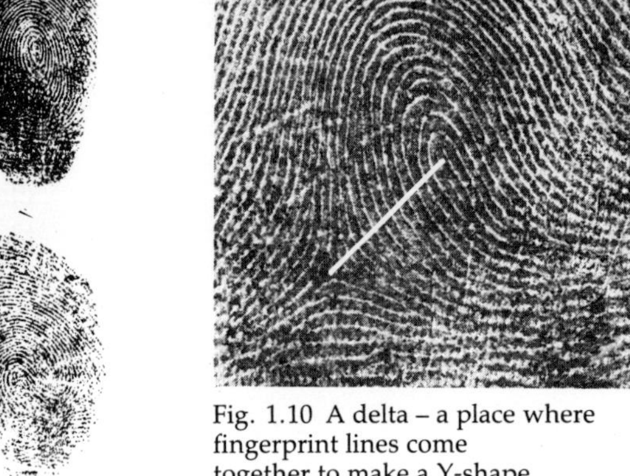

Fig. 1.10 A delta – a place where fingerprint lines come together to make a Y-shape

shown in Fig. 1.8. Try to name the four real fingerprints in Fig. 1.9 using the diagram. (Answers are on page 161.)

Fig. 1.10 shows one of the places where the fingerprint lines come together to make a 'Y' shape. These patterns are called *deltas*. They are important because scientists use them as measuring points. For instance, they do 'ridge counts' – counting the number of fingerprint lines between the middle of a loop or whorl and the nearest delta. The ridge count of the fingerprint in the diagram is 12.

Some features of fingerprints can be passed on from parents to children (inherited). They can give clues to the inheritance of some illnesses and diseases. Today's scientists are more interested in that, than in crime detection. They are also interested in fingerprint differences between humans and other animals. Do cats and dogs have fingerprints (pawprints!)? Try to find out.

Investigation 1.1 Taking your own fingerprints

You will need:
an ink pad
a magnifying glass

1 Roll your finger from side to side on an ink pad.
2 Make sure your finger is lightly coated with ink from the tip to the first joint.
3 Put a piece of white paper on the edge of a table. Roll your finger *once* from one side to the other. Take it off the paper carefully to avoid smudging.
4 Collect prints from all of your fingers.
5 Use a magnifying glass to examine your fingerprints. Name the patterns on each of them.
6 Choose a fingerprint which has at least one delta on it. Do a 'ridge count'. Compare your ridge count with your friends'.

Fig. 1.11 Taking fingerprints

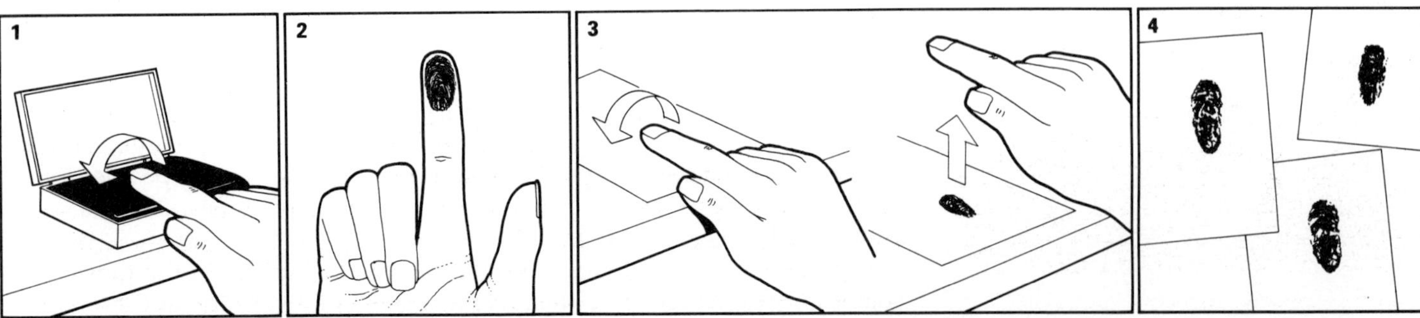

IN YOUR NOTEBOOK

✳ Write a heading: *Observing fingerprints*

✳ Imagine that you are writing a letter to a friend who was away when you looked at fingerprints. Tell your friend why fingerprints are special; how they are named; and what you did and found out in your investigation. Draw some pictures to make it all clearer.

Investigation 1.2 The detective game

You will need:
some plain postcards
an ink pad
talcum powder
a clean glass

1 Get together with four of your friends.
2 Make up a 'fingerprint file' with each person's name and prints on a plain postcard.
3 Choose someone to be 'the detective' holding the five cards.
4 Send the detective out of the room and choose one of the other four of you to be 'the criminal'. Get the criminal to take a firm grip on a clean glass. No one else must touch it!
5 Bring the detective back. See if he or she can 'spot the criminal' by comparing the prints on the glass with those on the cards. (It will help if you dust the prints on the glass with talcum powder. Blow off any spare.)
6 Wash the glass and give someone else a turn at being 'detective'.

1.4 Instrument makers

When scientists make their careful observations, it is as if they were picking up messages from their surroundings. They try to work out what the messages mean.

Some of the messages are very weak – light from a distant star or the heartbeat of an unborn baby, for instance. Some of the messages are difficult to pick up because we cannot sense them. Radio waves are like that.

For centuries, scientists have spent much of their working lives designing and making instruments to help them pick up the messages. Instruments like those in Figs 1.12, 1.13 and 1.14.

In 1590, in Holland, a young apprentice spectacle-maker held up two lenses, one behind the other. He was startled to find that he could see distant objects much more clearly. His master, Hans Lippershey, seized on the idea to make the first telescope. He tried to sell it to the government for use in warfare and it became one of the first military secrets! But the idea leaked out, and the great Italian scientist Galileo Galilei was the first to use a telescope to study the Moon, planets, and stars. Fig. 1.12 shows the largest lens telescope in the world at Yerkes observatory in the USA.

In 1910, a German scientist, Hans Geiger, was working at Manchester University. He was studying the invisible rays from some substances called 'radioactive materials'. People knew about the rays because they would blacken a photographic film put near them. Geiger needed an instrument which would tell him – quickly and easily – the direction and strength of the rays. It took him months to invent one which clicked when it picked the rays up. The stronger they were, the faster it clicked. Nowadays 'Geiger

Fig. 1.12 Telescope at Yerkes observatory

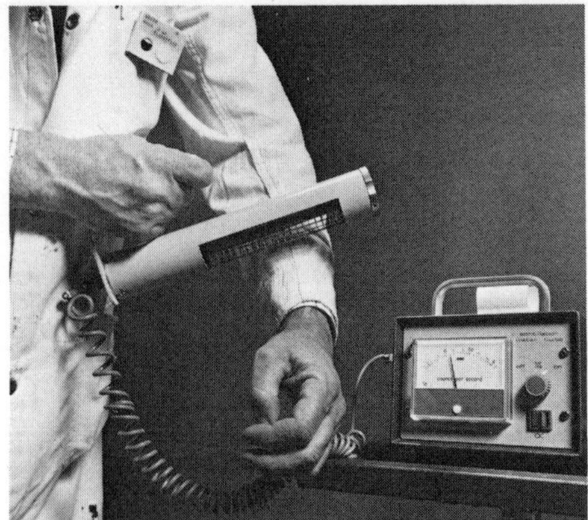

Fig. 1.13 Geiger counter in use. A radiation worker is checking his laboratory clothing to see if it has been contaminated by radioactive material.

Fig. 1.14 Jodrell Bank radio telescope

counters' are used all over the world to detect and measure the strength of atomic radiation.

In 1931, at the Bell Telephone labs in the USA, 26 year old Karl Jansky was studying the 'crackle' – 'static' – which was a nuisance in long distance telephone calls and radio messages. He discovered a new kind of static which had a pattern to it. It changed in step with the movement of stars across the sky. He realised that it was caused by radio waves from the stars. Hearing about this, another American, Grote Reber, built the first radio telescope to study these 'messages'. Today, there are radio telescopes all over world. Fig. 1.14 shows the one at Jodrell Bank near Manchester.

You can divide scientific instruments into two kinds:

1 Those that strengthen or concentrate a signal that we *can* sense. For instance they might make a signal easier to see or hear.
2 Those that change a signal which we *cannot* sense into one that we can. For instance, they might make an invisible signal produce a sound or a picture.

Look again at the instruments in the pictures. Try to say whether they are type 1 or type 2.

Galileo's telescope collected and concentrated the light from the planets and stars, making them easier to study. A light telescope is a type 1 instrument. A radio telescope is type 2 – it changes radio signals into sounds or into electrical signals which can be used to draw pictures.

IN YOUR NOTEBOOK

✳ Write a heading: *Scientists' instruments*

✳ List as many different scientists' instruments as you can. Next to each,
write a word or two to show what the instrument does. For example:

Instrument	Detects, collects, or measures
telescope	light from distant objects
Geiger counter	atomic radiation
metre scale	length

Try to get at least ten instruments. Compare your list with your friends'.

✳ Try to find out what these instruments do and how they work: *pedometer,
hydrometer, pyrometer*.

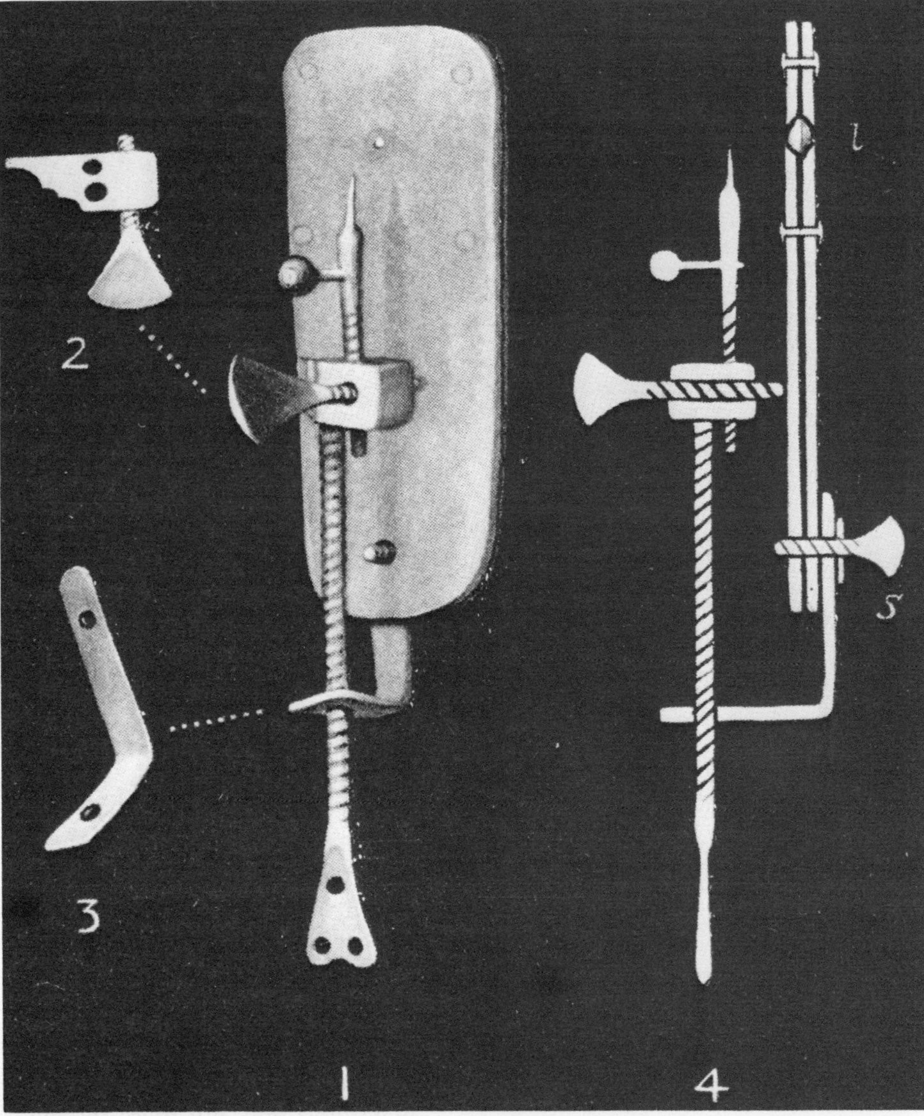

Fig. 1.15 A reconstruction of Leeuwenhoek's microscope. 1, 2 and 3 show the parts of the
microscope itself, and 4 is a side view drawing of it fitted together.

You have probably used one of the most useful scientific instruments – the microscope. Follow the instructions below to make one for yourself. It is a model of the instrument in Fig. 1.15, which was made more than 300 years ago.

This microscope has a tiny glass bead for a lens. To start with you have to make the lens. You need a piece of glass tubing 6 mm wide, 25 cm long.

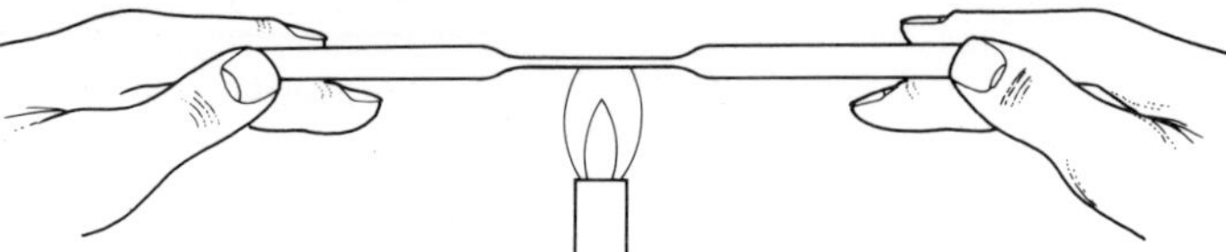

1 Heat the tube in a bunsen flame and draw it out.

2 Ask your teacher to cut it like this.

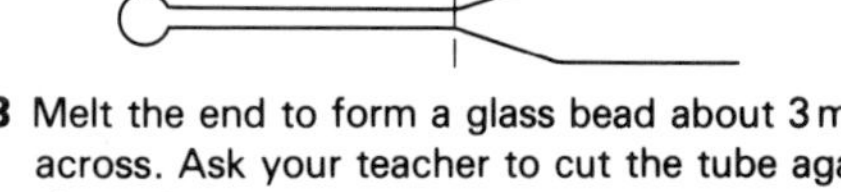

3 Melt the end to form a glass bead about 3 mm across. Ask your teacher to cut the tube again along the dotted line. Try to make a bead without air bubbles.

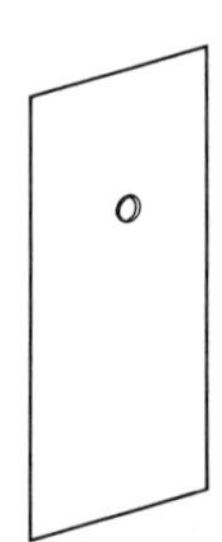

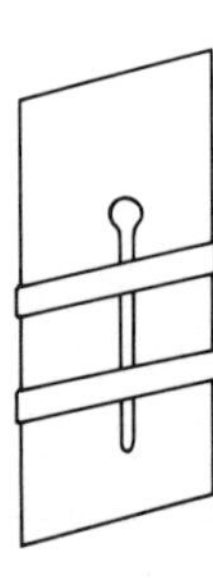

4 Take a piece of card 5 cm × 10 cm. Make a little hole in it. The hole should be a little smaller than your bead lens.

5 Fasten your bead lens just over the hole in the card. Use two rubber bands.

6 Trap a microscope slide between the lens and the hole. Look at the slide through the lens. Have your eye very close to it. Put a 60 W pearl bulb behind it.

Fig. 1.16 Constructing a model microscope

A microscope 'slide' is a thin oblong piece of glass with a specimen – such as a smear of blood – on it.

Your teacher will lend you a slide or help you to make one of your own.

1.5 Inventors of new words

Scientists take great care in describing the things they observe. They have invented many special words to say what things are like. You could almost say that they had a language of their own. Amongst scientists, these special words mean much more than ordinary words. Often they are a kind of shorthand.

Can you say what things are being described in these two puzzle passages? (Answers are on page 161.)

Material 1. A translucent liquid, off-white in colour. It is less dense than water and milk but more viscous than both. It contains much fat.

Object 2. A brown disc-shaped object with a rough surface. It is approximately 6 cm in diameter and 4 to 5 mm thick. It is relatively hard, but brittle.

It will help if you know the meanings of some of the words in the passages. For instance:

translucent: 'lets light through, but cannot be seen through'.
less dense: 'the same volume weighs less' (also, less dense liquids float on more dense ones).
more viscous: 'thicker, less runny than'.
hard: 'difficult to change the shape of'.
brittle: 'breaks easily if bent'.

Scientists use many words such as density, viscosity, and hardness to say what materials are like. They tell how the materials will behave under different conditions. They describe the *properties* of the materials.

Here is a list of 'property words'. The words on the right are supposed to be the opposites of those on the left but they have been jumbled up. Try to sort them out. Have a word with your teacher if you get stuck. (Answers are on page 161.)

A	brittle	1	biodegradable
B	conducting	2	flexible
C	hard	3	inert
D	heavy	4	insulating
E	reactive	5	light
F	rigid	6	soft
G	opaque	7	elastic
H	rotproof	8	transparent
I	strong	9	weak

IN YOUR NOTEBOOK

✱ Write a heading: *Property words*

Everyday object	Properties of material	Everyday object	Properties of material
aircraft wing		greenhouse side	
breakable cover	transparent rigid brittle	fishing rod	
drinks container		motorway crash barrier	

Fig. 1.17

✱ Copy Fig. 1.17, and try to complete the empty columns. First, you have to think about the material used to make the 'everyday object'. Then you have to choose at least two suitable properties for the material. For example, the material for a breakable fire alarm cover needs to be transparent (can be seen through), rigid (holds itself up) and brittle (breaks easily if bent).

1.6 Measurers

In the ginger biscuit puzzle, you probably found it useful to know the size of the disc in centimetres – a glance at a ruler and you knew just how big the mysterious object was. Scientists like to use numbers when they are describing things or happenings. For instance, they like to say *how much* of a particular property a thing has. That usually means they have to use some kind of measuring instrument.

A good measuring instrument has to do two things:

1 It has to enable you to compare two things so that you can say which has more of a certain property and which has less. For instance, you can use your foot as a measuring instrument to settle an argument about which of your friends has the biggest feet.

2 It has to give you a number to show how much of the property an object has.

A good way to understand the ideas behind measuring instruments is to make one yourself and study the way it works. You may have seen this instrument before. Make it now and study the way it works.

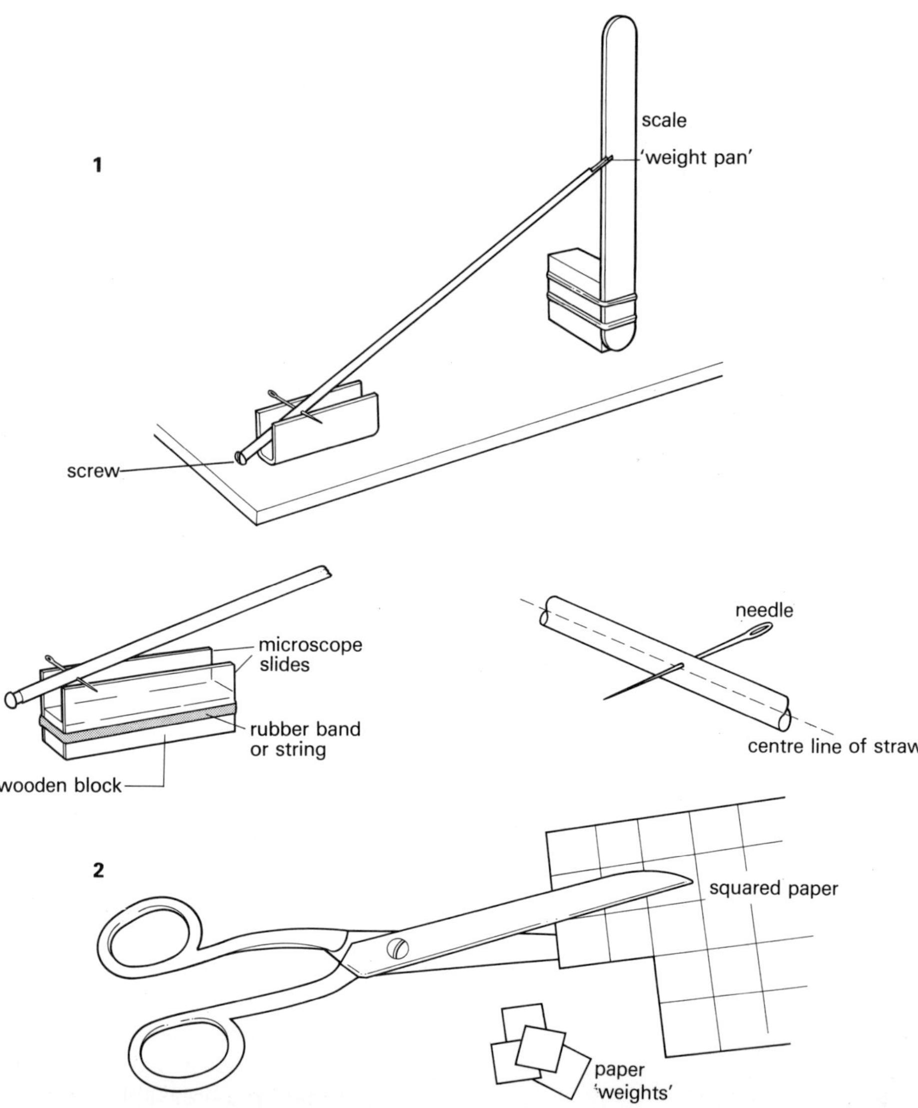

Fig. 1.18 Making a milk straw balance. The 'knife edge' may be made from a piece of metal U-channel or from microscope slides and a wooden block as shown.

Investigation 1.3 Making and using a milk straw balance

You will need:

milk straw	a supply of rice or small beads
needle	wooden strip
small, self-tapping screw	small wooden block
piece of metal U-channel	rubber bands
squared paper	scissors

1 Set up the milk straw balance.
2 Make some small 'weights' by cutting out pieces from the squared paper.
3 Use the paper 'weight units' to test the balance:
   Put one unit in the 'weight pan'. Mark the 'scale' to show where the end of the straw comes.
   Repeat with two 'units', then three and four.
4 Try to answer these questions:
   (a) Why is it useful to be able to wind the screw in a little, or unwind it?
   (b) What difference does it make if you move the needle closer to the screw?
   (c) What difference does the position of the needle make? What happens if you pull it out and move it further from the centre line of the straw?
   (d) Does the balance deflect twice as far with 2 units as it does with one; three times as far with 3 units, and so on? In scientific words, does it have a *uniform scale*?
   (e) How much does a rice grain weigh in your 'weight units'?
   (f) Is your balance *sensitive* enough to show the difference in weight between rice grains? Explain how you get your answer.
   (g) If you have never seen a 1 gram mass, ask your teacher to show you one. How could you find out how many of your 'units' weigh the same as a 1 gram mass? What other equipment would you need? (*Hint*: you can use as much squared paper as you like.)

Use your balance to weigh some other small things.

IN YOUR NOTEBOOK

✳ Write a heading: *A milk straw balance*

✳ Draw a diagram of the balance.

✳ Write out questions (a) to (g). Write your answers underneath.

1.7   Pattern makers

Scientists are not content simply to observe and write down their observations and measurements. If they can, they like to link them all together. The first step in doing this is to group things and happenings together so they have something in common. It is rather like stamp collecting. First you collect your observations, facts, and measurements (your *data*), then you arrange them in interesting ways.

You can arrange the stamps in Fig. 1.19 in many different ways. Suppose you had to pick out groups of stamps with something in common. How many groups could you pick? Try it. Try to say what *all* the stamps have in common.

Fig. 1.19 How many groups can you make from this stamp collection?

Fig. 1.20 A fisherman weighing a carp – a giant goldfish!

Stamp collectors enjoy arranging their stamps in groups. They get a better idea of what they have, and they can concentrate on one group at a time.

Sometimes the pattern of a group leads to a search for new additions. For instance, suppose one of the groups was: 'all stamps with pictures of animals'. The Zambian stamp showing a bird looks as if it might belong to a set. A stamp dealer says it does – he can supply the others – and the collection gets bigger!

The same kinds of things happen when scientists organise their data into groups. To start with, life is less confusing. They can concentrate on one group at a time. Then they get a better idea of the gaps in their collection – and where to look for new observations and measurements to fill them.

When scientists write down what a group of things or happenings has in common, they call the statements *patterns*.

'*All the stamps on the page have perforations round their edges*' is a pattern.

In the next few pages you will be looking at examples of scientists' patterns. Start with one that is 300 years old . . .

Suppose you asked the fisherman what he was using to weigh his fish. He would probably say: 'a spring balance'. A scientist would say he was using a *forcemeter* or a *newtonmeter* or a *spring dynamometer*.

1.8 An elastic pattern

The forcemeter makes use of an important scientific pattern. Have you come across it? Can you say what it is?

The pattern was discovered 300 years ago by a famous English scientist, Robert Hooke. It says that you need twice the force to stretch a spring twice as far, three times the force to stretch it three times as far, and so on. In scientific words: *the force needed to extend a spring is proportional to the amount of extension*. This is usually called *Hooke's law*.

You can show the pattern with a graph like the one in Fig. 1.21.

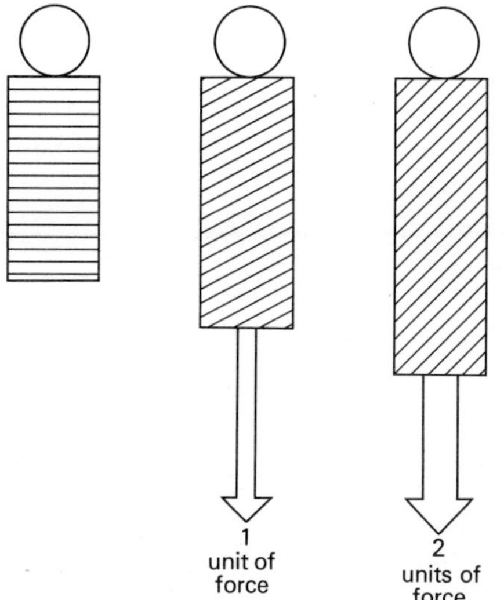
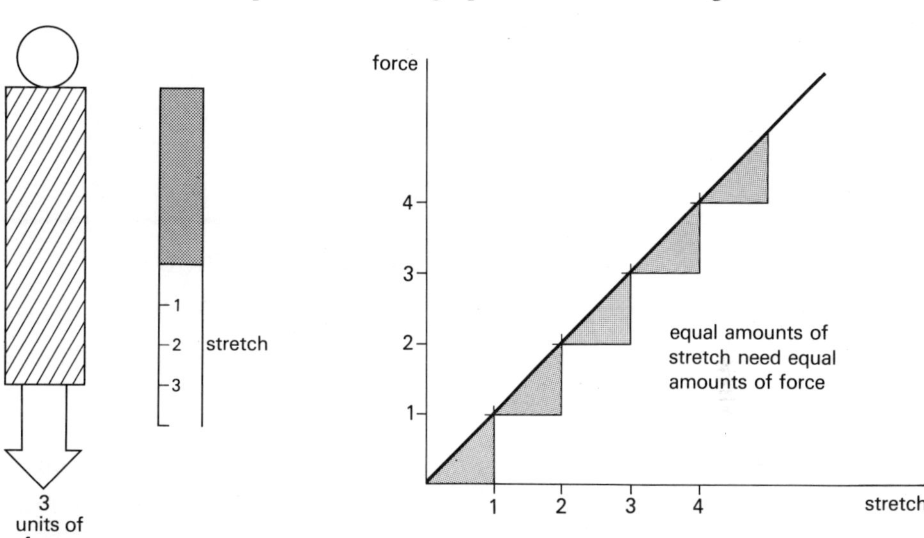

Fig. 1.21 Hooke's law

Hooke's law is a pattern because it tells us what all springs have in common. Long springs, short springs, thin springs, fat springs, iron springs, and copper springs, all follow the same stretching pattern. In fact, since Hooke's day, scientists have discovered that many other things fit into the pattern, for example a length of catapult elastic.

There are plenty of forcemeters in a school lab, but most of them measure forces of tens or hundreds of newtons. Suppose you wanted to measure a very small force, for instance, the driving force of a battery powered toy car, or a model steam engine. You would need a much more sensitive instrument. Here are some instructions for making your own sensitive forcemeter.

Investigation 1.4 Making and using a sensitive forcemeter

You will need:
shirring elastic ('shirring silk')
thread
a small washer or ring
a 10 N forcemeter (for a short time only)
model electric or clockwork cars, boats or trains;
    perhaps even a model steam engine
cardboard strip 30 cm × 4 cm
graph paper
transparent sticky tape
scissors

1 Cut a notch in the strip of cardboard then a small slit at the bottom of the notch.

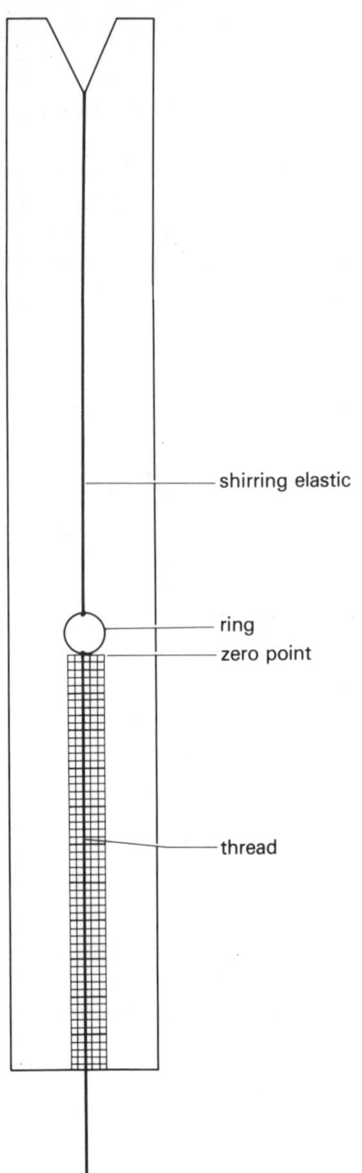

Fig. 1.22 Home-made forcemeter

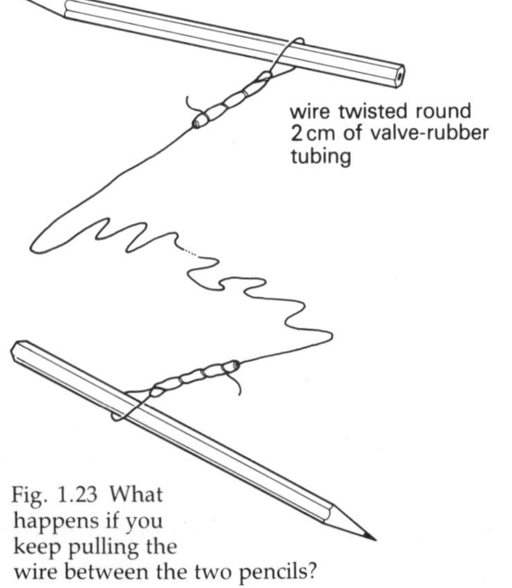

Fig. 1.23 What
happens if you
keep pulling the
wire between the two pencils?

2 Tie a big knot in one end of the piece of shirring elastic and fix the elastic in the slit with the knot at the back.

3 Tie the washer (or ring) to the other end of the elastic so that it lies on the card as shown in the diagram.

4 Cut a strip from the graph paper and tape it to the card to make a scale running from the ring to the end of the card.

5 Pull on the ring with the 10 N forcemeter. Find how far the ring moves for a force of 1 N. Mark a figure 1 at the right place on the graph paper. Then put another mark twice as far away from the zero point and mark it 2 and so on.

Marking a blank scale in standard units (such as newtons) is called *calibrating* the scale. Hooke's law makes it easy to calibrate your scale because: 'twice as much force gives twice as much stretch' . . . and so on.

Now your home-made forcemeter is ready to measure the driving force of a model car. Try it out. Try to think of other small forces you could measure with it.

IN YOUR NOTEBOOK

✱ Write a heading: *Hooke's law and spring forcemeters*

✱ Write these statements, filling in the gaps:
Hooke's law is a pattern because it l _______ t _______ a group of things by saying that the force needed to stretch a spring is p _______ to the stretch.

✱ Copy this table and fill in the gaps:
(Use Hooke's law to work out what the figures should be. The 'units' are not given because the pattern fits whatever they are.)

Force in 'units'	Stretch in 'units'
—	2
2	
3	—

✱ Draw a diagram of your home-made forcemeter. Write down one or two measurements you made with it.

1.9 Stretching the pattern!

Springs and other materials will only stick to the Hooke's law pattern for so long. Pull on them too hard and you know what will happen!

The pattern is only part of the story of what happens when things are stretched. You can investigate the rest of the story with the apparatus shown in Fig. 1.23.

When you are ready, pull very gently on the two pencils. At first, the wire will hardly seem to stretch at all.

Pull a little harder and it feels 'springy'. If you relax it will spring back. It is fitting into the pattern.

Gradually pull harder and harder. *Do not jerk*. After a while you will feel the wire start to *run* like toffee and it will come apart into two pieces. Scientists would say you had 'taken the wire past its yield point'. A complete graph for the behaviour of the wire looks like Fig. 1.24.

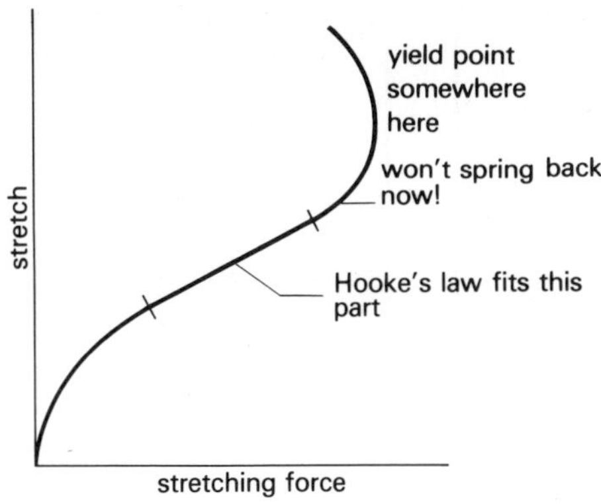

Fig. 1.24 Graph for the behaviour of the wire in Fig. 1.23

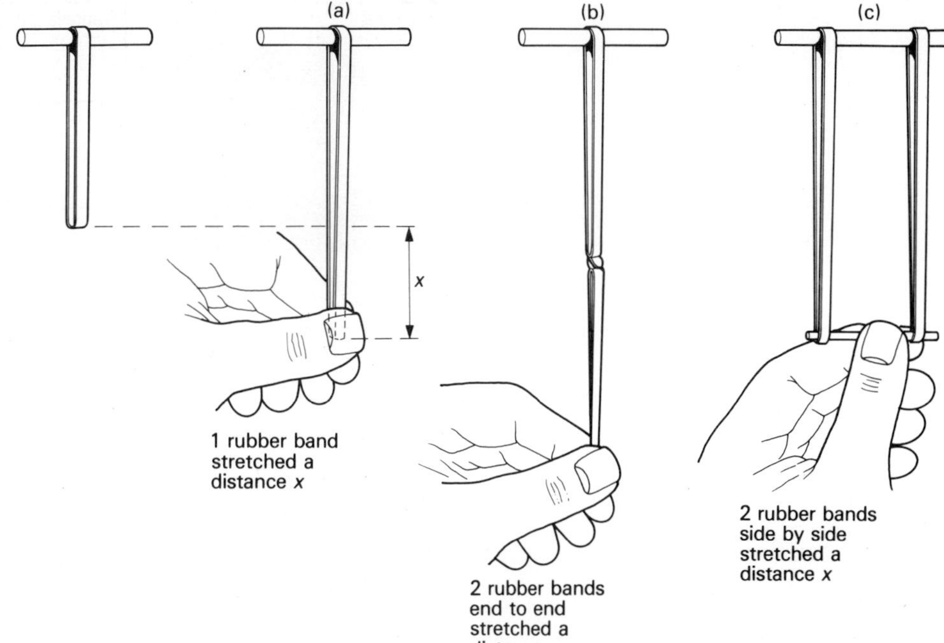

Fig. 1.25 Which experiment ((a), (b), or (c)) needs the greatest pull?

Finally, look at the puzzle in Fig. 1.25. Can you solve it? When you have decided what the answer is, design an experiment to test it.

1.10 Living patterns

Human beings love organising things into groups. Toddlers will happily spend hours sorting their bricks into colours. Hobbies such as stamp collecting and coin collecting are always popular. Can you think of other games and hobbies that involve sorting things into groups?

On page 12 you saw that scientists have good reasons for patterning their data – for sorting them into groups . . .

1 Life becomes less confusing! They can study one group at a time.
2 Old observations can be 'looked up' quickly. The grouping acts like a filing system.
3 New observations can be filed away quickly. They can be fitted into a pattern, sometimes causing it to be changed, sometimes not.
4 Patterning shows up 'gaps' in the data. New observations can be suggested.

There is a fifth reason. Scientists like to invent *explanations*. They want to be able to say *why* things are like they are – what might have *caused* them. Patterning often gives clues to this.

You can see all of this in action if you look at scientists sorting living things into groups. (The scientific word for any living thing is an *organism*. Scientists who are mainly interested in organisms are called *biologists*.)

Biologists have tried many different ways of sorting organisms. For instance, they have arranged them in groups which move in the same way, or in groups which live in the same place. At one time, plants were put in groups which showed their value for making medicines – headache cures here, laxatives there, and so on. The most popular method has turned out to be that invented by the Swedish scientist, Karl von Linné in 1735.

Here is an example of Karl's method – the *Linnaean classification system* – in operation.

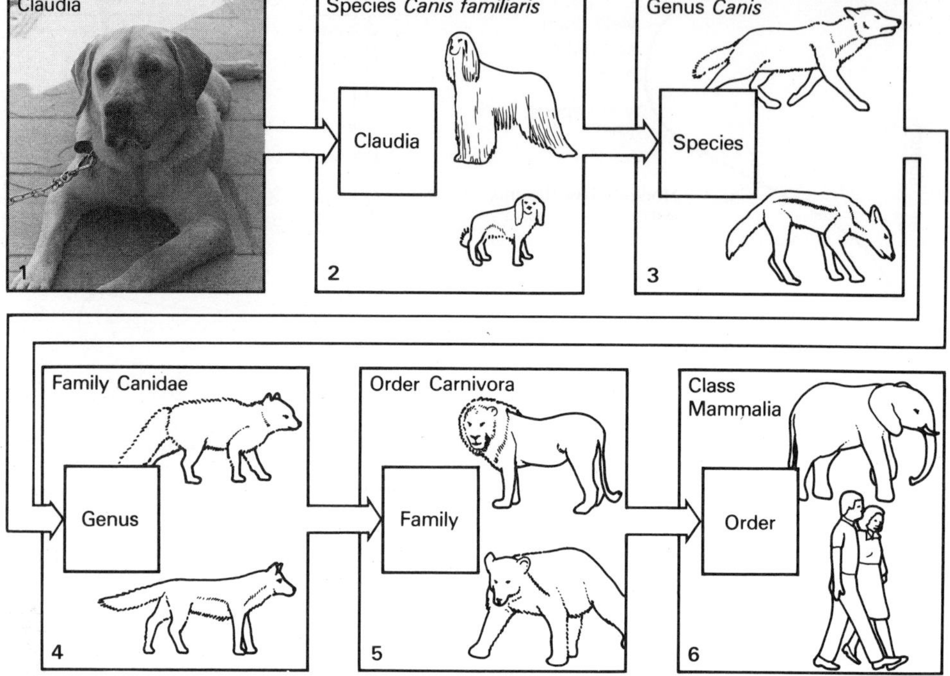

Fig. 1.26 Classifying 'pet dog'

1 This is Claudia. She is a female yellow labrador. She could breed with any other dog. A group of organisms which can breed with each other and produce offspring is called a *species*. Claudia is a member of a large group
2 of animals called the 'pet dog' species.
3 The group of pet dogs is part of a larger group of animals with several things in common. These animals have long heads, pointed ears, 42 teeth, and claws which they cannot pull back into sheaths in their paws.

Such a group, made up of similar *species* groups, is called a *genus*. Dogs belong to the genus *Canis*.

So Claudia might be classified as a *Canis familiaris* (pet dog!). This system of giving organisms two names is called the binomial system. The first name (with a capital letter) shows the genus, the second shows the species. The names are rather like entries in a telephone directory – surname, or 'family name' first. The names are from the Latin.

The careful naming system avoids confusion. All scientists use the same name for the same organism. For instance, in different parts of the world, a dragonfly is called a horse-stinger, mosquito hawk, snake-feeder, snake-doctor, witch doctor and dragonfly. To all scientists, it is 'an insect of the order Odonta'.

4 The *Canis* genus and some genuses of fox-like animals are grouped together to make the *Canidae family*.
5 The *Canis* group is part of a larger group of animals linked together because they are meat-eaters. The meat-eating group is called the *order* of *carnivores*. There are many other genuses in the order of carnivores – the cats (*Felis*), for instance.
6 An even larger group containing several orders is called a *class*. The carnivores belong to the same class as we humans. Like us they are warm-blooded and breathe with lungs. Like us they give birth to live young which they feed with their own milk. Like us they belong to the class *Mammalia* – the mammals.

All the mammals have backbones, so they belong to an even larger group called the *Vertebrata*. A group containing many classes is called a *sub-phylum*.

You might think that this very large group – the sub-phylum – would include most of the animal species. No so! Only five species out of every hundred on the Earth have backbones!

There are many sub-phylums in the animal kingdom. In each sub-phylum there are many classes. In each class there are many orders. In each order there are many genuses. In each genus there are many species.

Classifying animals has helped scientists to work out how animals are related to each other; and how they might have developed from common ancestors. The groups have been changed many times to make them fit in with theory and to make them neater.

1.11 An interaction pattern

Many scientists spend their working lives studying living things. Others spend their lives studying chemical substances.

There are many scientific patterns to do with the things that happen when chemical substances come together. That is, to do with *chemical interactions* (or *reactions*).

In the investigations below you will be studying interactions between two families of chemical substances – *acids* and *carbonates*. You will be looking for a pattern which sums up everything that happens.

You will need:

a small test-tube	zinc carbonate
dilute hydrochloric, nitric and	cobalt carbonate
sulphuric acids in dropper bottles	nickel carbonate
copper carbonate	small pieces of limestone and marble

Investigation 1.5 Dilute hydrochloric acid and carbonates

Follow Fig. 1.28 using dilute hydrochloric acid. Make a note of anything you see.

Investigation 1.6 Dilute nitric acid and carbonates

Follow Fig. 1.28 again. This time use dilute nitric acid.

IN YOUR NOTEBOOK

✳ Write a heading: *An acid-carbonate interaction*

✳ Write these statements, filling in the gaps:

I added dilute ______ acid to four different carbonates. Each time . . .

Next I added dilute ______ acid to each of the carbonates in turn. The result was . . .

✳ Copy Fig. 1.27. Write a different word in each empty box to sum up what happened. (The arrow means 'gave' or 'produced'.) (The answers are on page 161.)

	+	CARBONATE	→	

Fig. 1.27

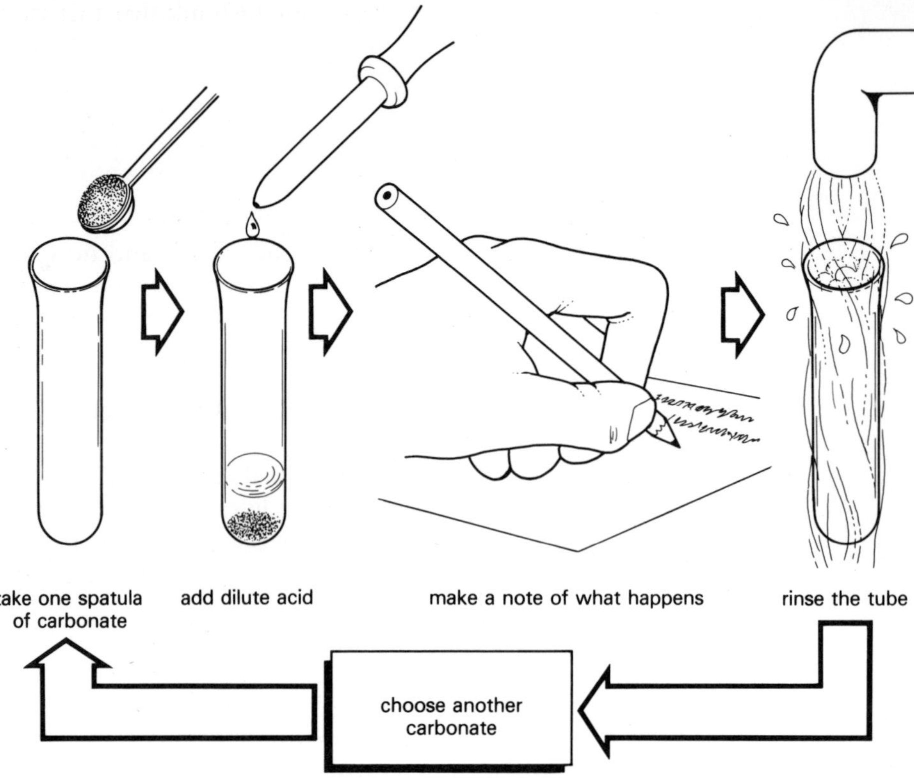

Fig. 1.28

The gas which is given off in the interaction is called *carbon dioxide*. This gas can be tested using a liquid called *limewater*. It is the only gas which changes limewater in a certain way. Look at Fig. 1.29 to see how to do the test. Try it.

IN YOUR NOTEBOOK

✳ Write this statement filling in the gaps:
   I tested the gas (carbon dioxide) with ______. *It turned* ______.

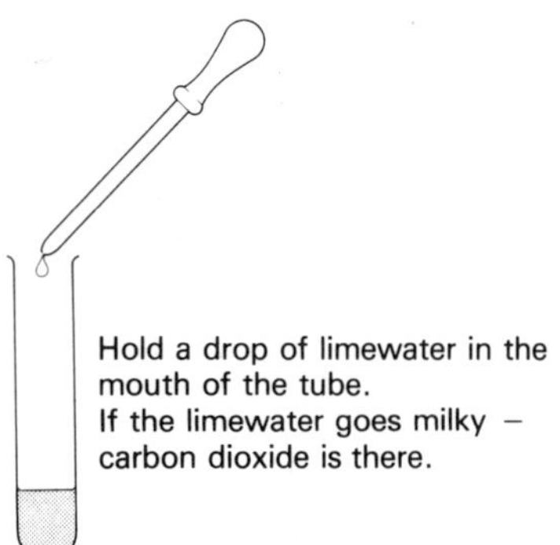

Fig. 1.29 Limewater test

Investigation 1.7 Limestone, marble and dilute acids

Use the two rocks, limestone and marble. They both contain large amounts of carbonate.

1 Put a spatulaful of limestone in your test-tube (Fig. 1.28). Add dilute hydrochloric acid to it. Make a note of what happens.
2 Do the test again, using marble instead of limestone.
3 Test limestone and then marble with dilute sulphuric acid.

IN YOUR NOTEBOOK

✳ Write a few words about the difference between the interaction of marble with sulphuric acid and the interaction of limestone with sulphuric acid.

✳ Do you think that the acid-carbonate pattern should be changed for sulphuric acid?
   Write down what you think.

Finally, read this passage and try to answer the questions.

Many old buildings are made out of limestone. If you look closely you can see that the stone has often been 'eaten away'. The marks and patterns of the damage suggest that rainwater might have caused much of it. Buildings made out of marble seem to do much better.

Remembering the acid-carbonate pattern, and the results of Investigation 1.7, what can you say about rainwater?

Can you see any other differences between limestone and marble? Could these have made a difference to the speed of 'eating away'?

When you think you have the answers, talk about them with your teacher.

1.12 Whole Earth patterns

So far, you have looked at patterns which link together groups of animals by describing the things they have in common. You have looked at a pattern which says what stretchy materials have in common; and another which says something about *most* acids and carbonates.

The patterns in Fig. 1.30 show what some *places* have in common – places scattered all over the Earth. The maps show these patterns. For instance, one shade of grey might show all the places with a certain kind of vegetation.

The idea is for you to study the patterns in the maps and to try to link them together to make 'patterns of patterns'. You might finish by being able to write statements which begin:

Most places with ______ vegetation seem to have . . .

You will need another map of your own – just an outline – traced on to a sheet of plain paper.

IN YOUR NOTEBOOK

✳ Write a heading: *Whole Earth patterns*

✳ Stick your map in your book.

✳ Mark ten places on the map – anywhere you like but spread them out – and label them A–J.

✳ Make a table like this with a row for each place:

Place	Vegetation	Temperature	Rainfall
A			
B			
C			
⋮			

✳ Use the information from the maps to complete the table.

✳ Try to write some statements like these:
All places with ______ type of vegetation seem to have ______ temperatures and ______ rainfall.

✳ Make a note of any places which did not seem to fit into a pattern. Talk about them with your teacher.

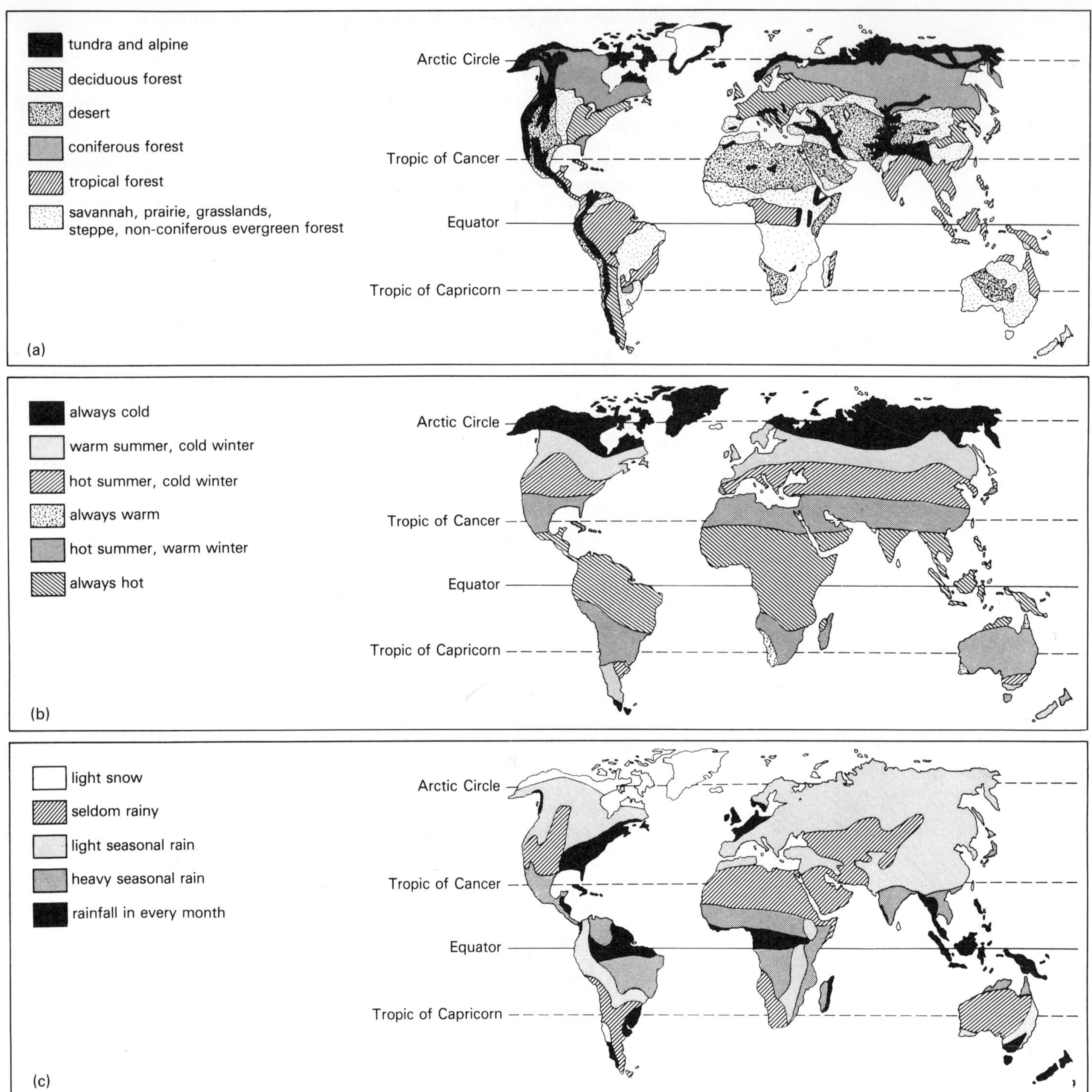

Fig. 1.30 (a) Vegetation of the world
         (b) Temperatures in the world
         (c) Rainfall in the world

Important note

It is tempting to say that because all places (or most places) with a certain kind of vegetation have a certain kind of climate, the climate *caused* the vegetation. Scientists are always careful about statements like that. It *may* be true that vegetation A was caused by climate C, but it may not! Both might have been caused by other factors such as:

1 the height of the ground above sea level; **2** the activities of humans; **3** the activities of animals.

Can you explain how these three factors might change the vegetation? Try it.

1.13 Weather patterns

A hundred and sixty years ago, a boy called Francis Galton was born in Birmingham. He was to become one of the most remarkable scientists of all time – though less famous than his 13 year old cousin Charles Darwin. By the time he was three Francis could read almost as well as an adult. By the time he was 25 he had studied to be a doctor, decided that he didn't want to practise medicine, and had set off to explore Africa!

During his lifetime he worked out how to identify people from their fingerprints, invented the high-pitched whistle which dogs can hear but we cannot, and was the first person to study identical twins. In 1863 he wrote a book which founded today's method of drawing weather maps. Fig. 1.31 shows how today's weather maps are produced. You need to know the

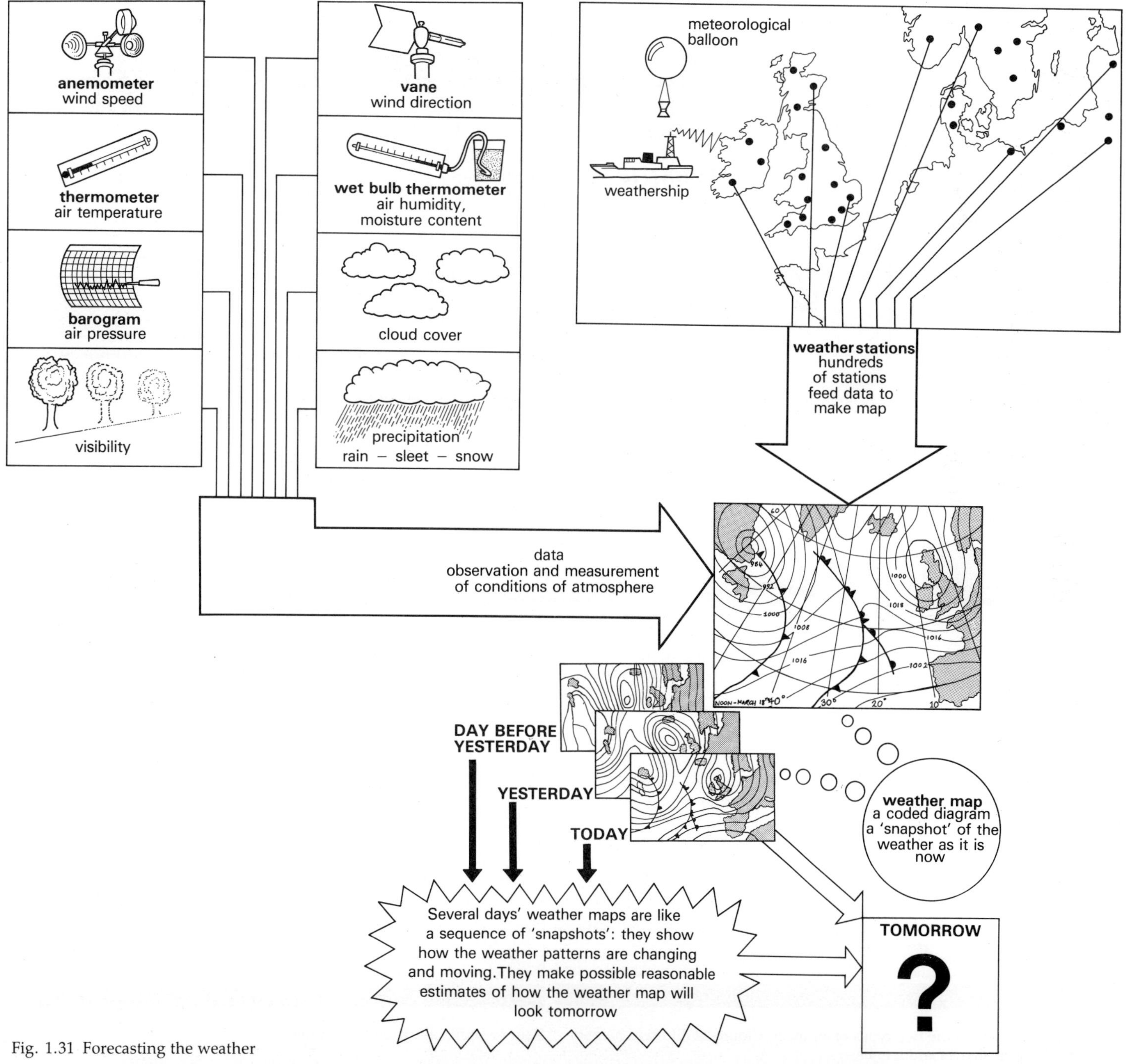

Fig. 1.31 Forecasting the weather

pressure of the atmosphere at all the places on the map. You draw lines to join up the places with the same pressure. The lines are called *isobars*. You can see them, marked with the pressure, on the weather map in Fig 1.31.

Finding patterns in the changing shapes of isobars is the secret of weather forecasting. You can see some of these shapes on the weather map. The whole diagram shows what it takes to make 'today's forecast'.

1.14 Local forecasts

The weather scientists – *meteorologists* – compare the daily 'snapshot' weather maps and try to 'predict' what the next one will be like. There's no certainty in making these predictions. The scientists know that the pattern can change – unpredictably – at any time.

The most useful features of the maps are the lines marked with spikes and bumps; and the circular isobar patterns marked 'LOW' and 'HIGH'. There are two lows on the map in Fig. 1.31. The lines with spikes and bumps are called *fronts*. They show the edges of great moving masses of cold and warm air. Fronts are explained in Fig. 1.32.

What is a front?

Fig. 1.32

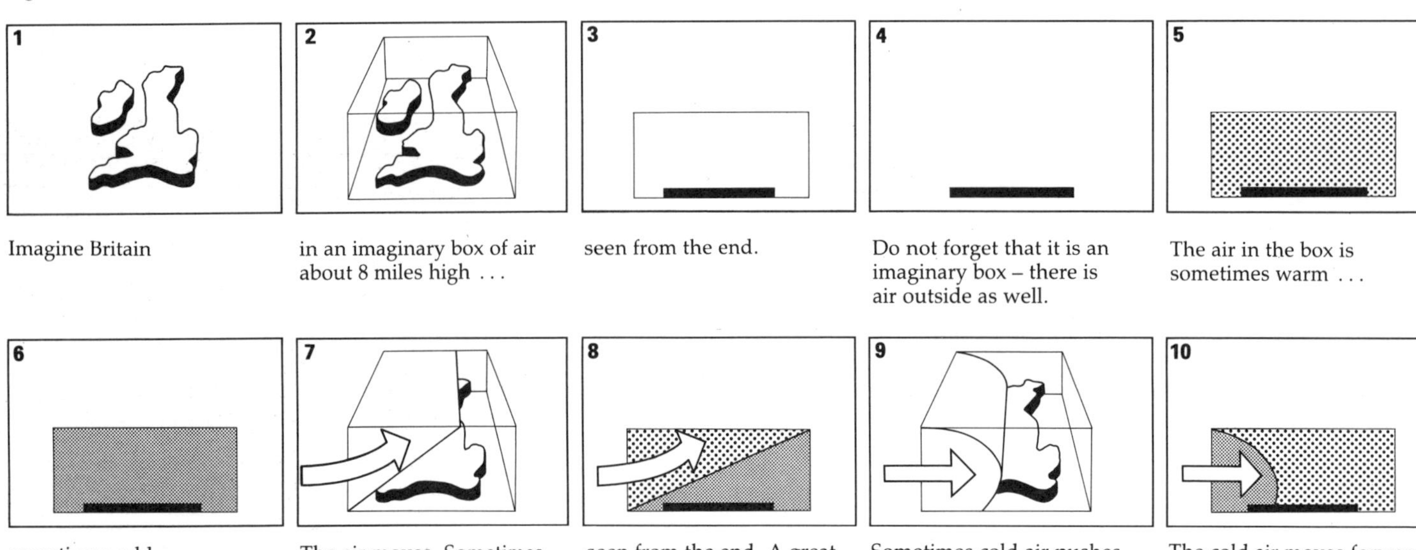

1 Imagine Britain

2 in an imaginary box of air about 8 miles high . . .

3 seen from the end.

4 Do not forget that it is an imaginary box – there is air outside as well.

5 The air in the box is sometimes warm . . .

6 sometimes cold.

7 The air moves. Sometimes warm air pushes cold . . . like this . . .

8 seen from the end. A great wedge of warm air riding upwards over cold air beneath – *a warm front*.

9 Sometimes cold air pushes warm.

10 The cold air moves forward as a great rolling mass – *a cold front*.

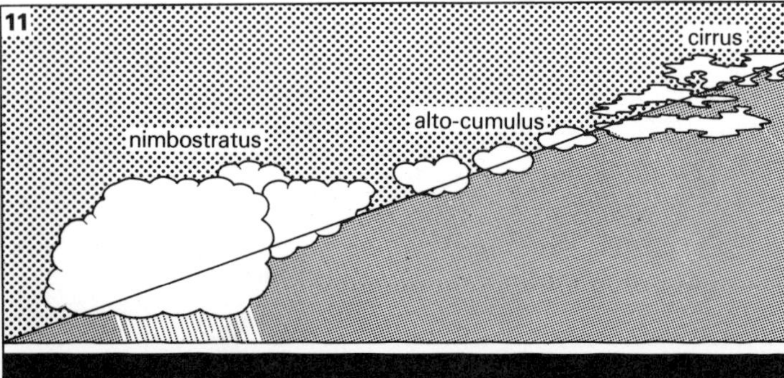

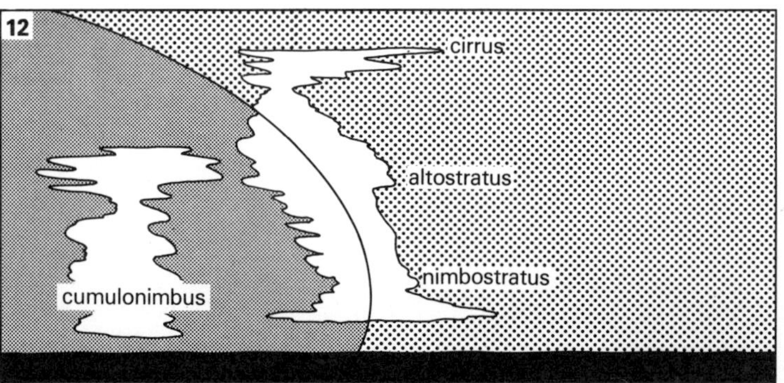

Clouds form where the two types of air meet. Clouds at different heights have different shapes and names.

Rain or snow falls from the clouds.

1.15 Forecasting

If the weather map tells you that a front is coming your way you can forecast rain.

The first signs of the front are high clouds. The 'mare's tail' *cirrus* clouds of a warm front are made of ice. They could be 8 miles high and 20 hours ahead of the front – the bottom of the wedge – at ground level. The great flattened head of a *cumulonimbus* thunder cloud could be 4 miles high and lurking 5 hours behind a cold front.

You can tell when the front reaches you at ground level. The wind changes direction as the front passes. If you are facing the wind it will quite suddenly veer round to your right.

If you have a barometer you can check the pressure too. It will rise as the edge of a cold front passes, or fall for a warm front.

IN YOUR NOTEBOOK

✷ Collect weather maps for at least four days running. There are good ones every day in *The Guardian* and *Daily Telegraph* newspapers. Stick them into your book.

✷ Keep a record of the weather at your home over the same period. Write down the times of any showers or heavy rain.

✷ Write a short account of the way that the weather changed. Try to link it with the patterns in the weather maps.

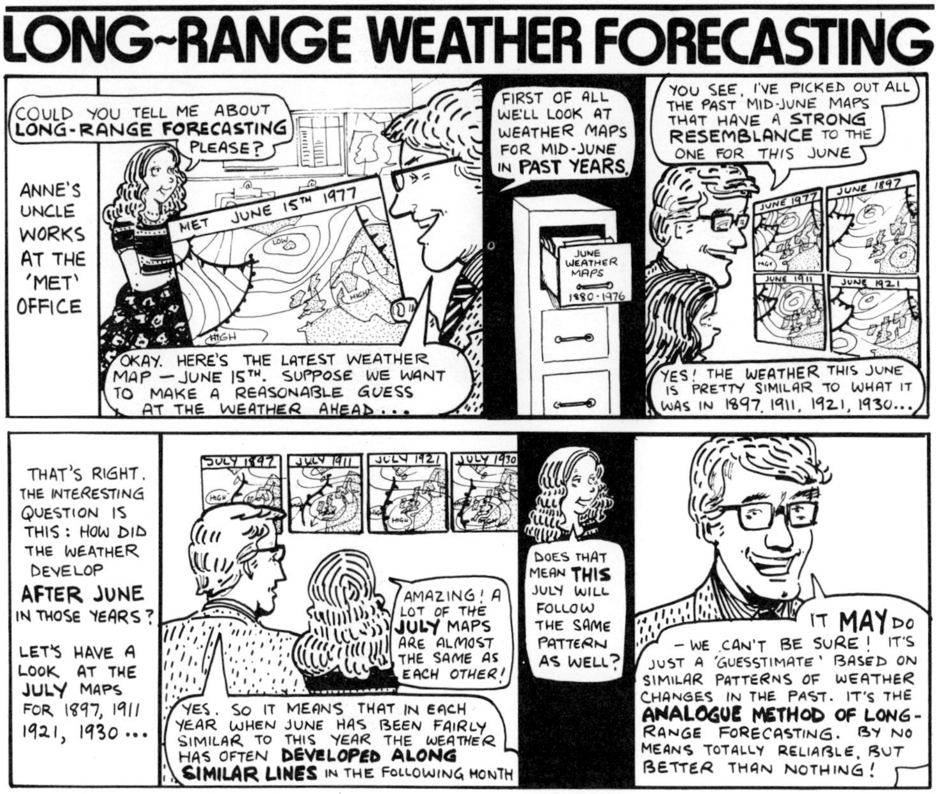

Fig. 1.33

As well as making local daily forecasts, meteorologists make long-range forecasts. These are a very good example of patterns being used to make predictions. You can see how they are made in Fig. 1.33.

Over the last few years, meteorologists have been helped by the pictures from satellites orbiting over the Earth. The photograph in Fig. 1.34 was taken on one of the days of the Ascot races in 1980. The great swirl of cloud to the north-west of Britain is a giant 'LOW' area. The bands of cloud swinging out from it are fronts. You can see them marked on the synoptic chart alongside.

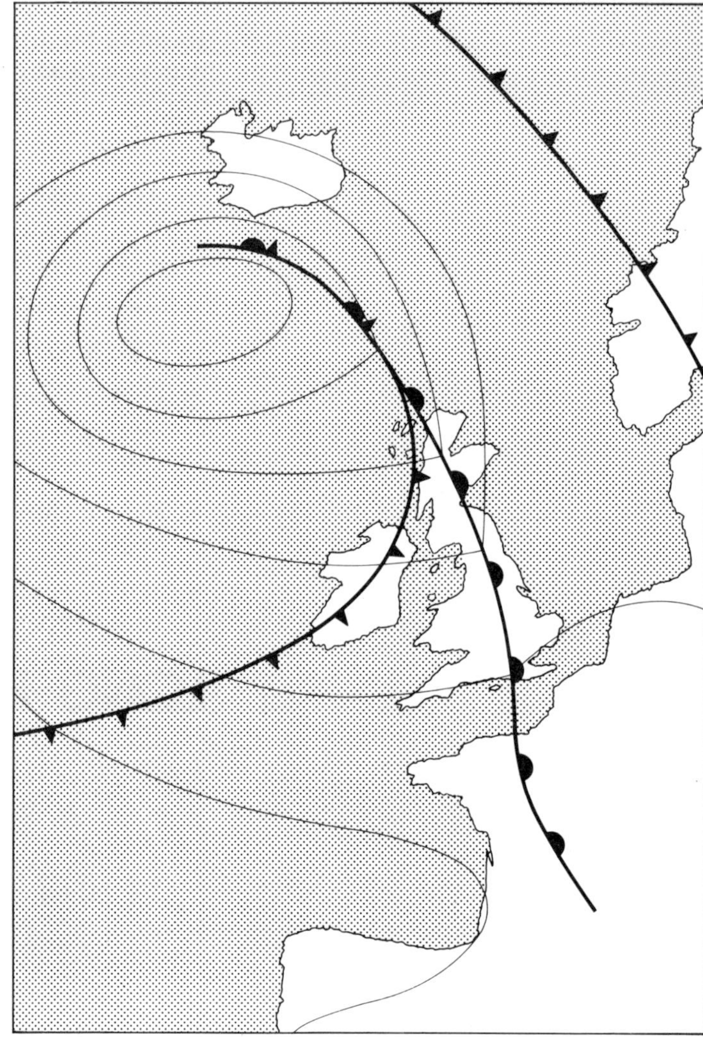

Fig. 1.34 Satellite picture taken over the British Isles with the synoptic chart alongside

Study the photograph and the chart and try to answer these questions:

1 What was the weather like at Ascot when the picture was taken? (You will probably need an atlas to find Ascot on the British Isles.)
2 What was the weather like in Cornwall? (Answers are on page 161.)

IN YOUR NOTEBOOK

∗ Write a heading: *Who needs weather forecasts?*

∗ Try to answer these questions:

1 Who needs to know what the weather will be like in advance? Why do they need to know?
2 How can weather forecasts save lives?

Checkout

Keywords

calibrating (a scale)
chemical interaction
classification
data
explanation
fact
instrument
isobar
measurement
meter

observation
organism
pattern
prediction
property
sensitivity (of an instrument)
signal
species
uniform scale
weather front

Patterns

1 Scientists try to make all their observations *double* observations – so that they have at least two things or happenings to compare.

2 Scientists try to arrange all their data in groups with at least one common feature. This 'patterning' makes the data easier to study; suggests where new observations might be found; and makes new data easier to handle and explain.

3 Scientists' instruments may be divided into two groups:
  (a) Those which strengthen signals we can sense.
  (b) Those which change signals we cannot sense into those we can.
  Both types of instrument may be modified to *measure* the signals.

4 There is a pattern which connects the force needed to stretch any spring with the amount of stretch the force causes.

5 Most acids and carbonates are connected by the pattern

ACID + CARBONATE → CARBON DIOXIDE GAS

6 There are patterns which connect the vegetation at various places on the Earth with the climate at the same places.

7 Patterns can be seen in the shapes of isobars drawn on weather maps. The patterns can be used to forecast changes in the weather.

True or false?

1 A scientific pattern does not have to be complete to be useful.

2 Both the following instruments are designed to change signals we cannot sense into those we can:
  (a) thermometer;
  (b) forcemeter.

3 The main reason scientists invent new words is to keep their discoveries secret.

4 A material can be both hard and brittle.

5 A spring is 15 cm long with a force of 5 N pulling on it; and 20 cm long with a force of 10 N. With no force pulling on it, the spring will probably be 10 cm long.

6 All animals in the same *genus* can breed with each other and produce offspring.

7 Most vertebrates are mammals.

8 All carbonates produce carbon dioxide gas with hydrochloric acid.

9 The shape of a cloud gives a good indication of its height.

10 Some parts of the Earth's atmosphere have no weather.

Problems

1 *Read the passage and try to answer the questions.*

About 60 years ago, a young scientist called Alan Griffith was working at the Royal Aircraft Establishment labs at Farnborough. He was looking for answers to the questions: How strong can a material be? What is the greatest possible strength of a material? What would a super-strong material be like?

He had used Hooke's law, some other patterns, and a theory about the way materials are made up to calculate some answers.

His answers for the greatest possible strength seemed far too high. When actual materials were tested they were far weaker than the calculations said they should be. But it was difficult to draw good conclusions because test data was scarce. So Griffith decided to collect more data by measuring the strengths of very thin threads – *fibres* – of a material.

He chose glass as his material – making the threads by pulling apart the ends of heated glass rods. He stretched each of the threads and measured the force needed to break it. His results are shown in Fig. 1.35.

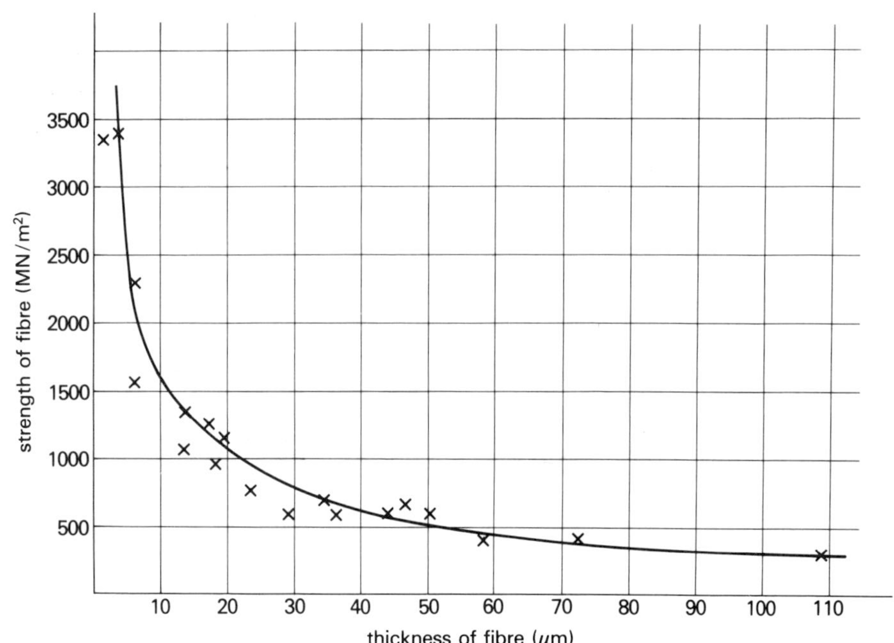

Fig. 1.35

(The units of strength are millions of newtons – meganewtons – per square metre. The thickness of the threads is measured in millionths of a metre – micrometres.)

Griffith was pleased with his results. They showed that glass fibres *could* be as strong as he had calculated. (10 000 MN/m^2)

The results encouraged other scientists to study materials in the form of fibres. Today these materials are of great importance in industry and everyday life.

(a) Why do you think Griffith chose glass for his experiments instead of a metal such as steel?

(b) What pattern does the graph show? Write it down.

(c) How did the graph show that real fibres could be as strong as Griffith had calculated (10 000 MN/m^2)?

(d) How many things made out of glass fibre can you name? Try to write down at least five.

2 *The weather maps in Fig. 1.36 were drawn at lunchtime on four December days. Compare the maps with the descriptions of the weather in London on the same four days. Then try to answer the questions.*

Wednesday 17 December
A fairly mild day, colder in the evening. Light rain fell steadily all day with some heavy showers around 3 pm. The wind was from the south-west and quite strong.

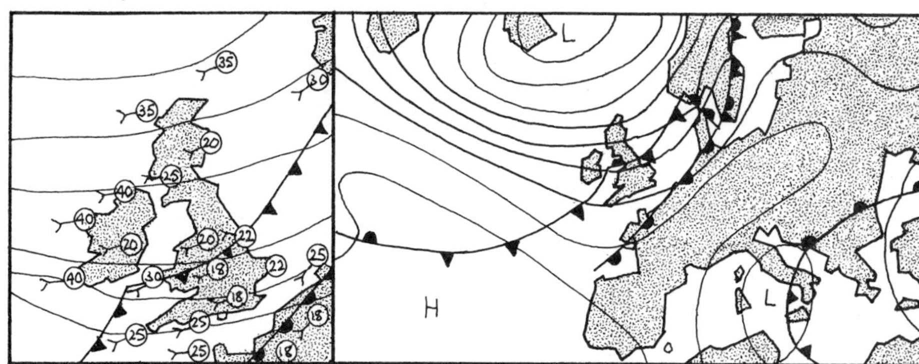

Thursday 18 December
A cold day. It was dry with long sunny periods. A strong wind blew from the west.

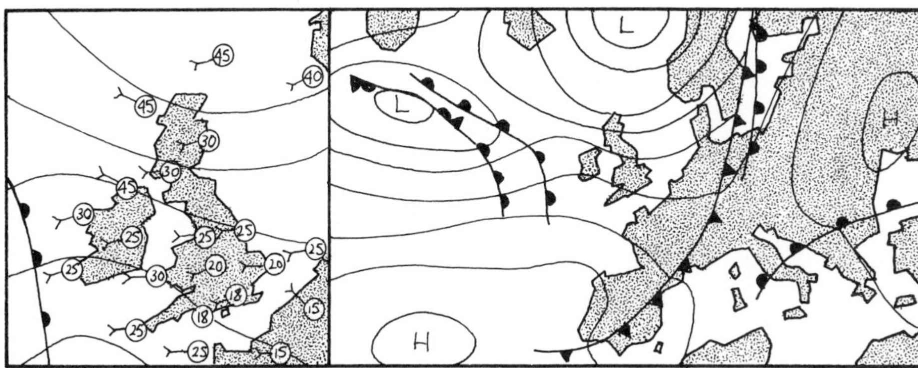

Friday 19 December
Another cold day but becoming much milder out of the wind in the evening. It was overcast but dry in the morning and rained fairly heavily after lunch and into the evening. The wind started north-easterly but changed to south-westerly at teatime.

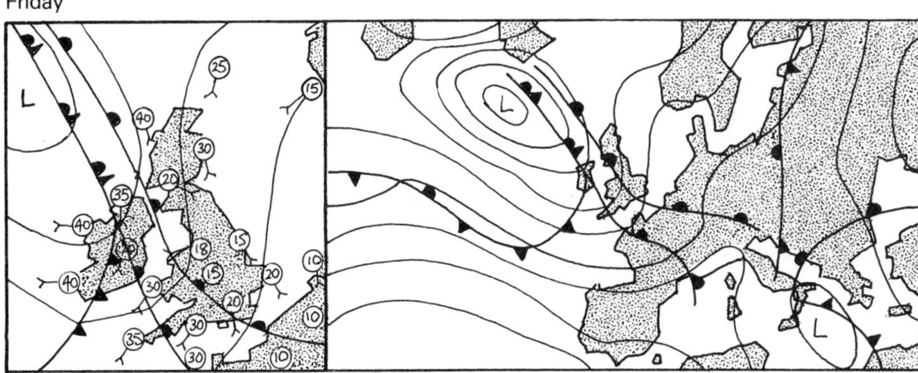

Saturday 20 December (Morning only)
Fairly mild, with heavy showers of rain and a strong north-westerly wind.

Fig. 1.36

(a) How do you explain the weather on the four days? Take each day in turn, and link the movement of the weather fronts and the wind directions, to the weather in London.

(b) Can you see a pattern which links the wind directions to the shapes of the isobars on the maps? Write the pattern down.

(c) You are only given the weather for Saturday morning in London. What would your forecast for the afternoon have been, if you had seen the map at lunchtime?

2 Earth

2.1 Volcano!

Fig. 2.1 A volcano on the west coast of Mexico erupting. What are the light patches on the left of the crater? What are the ridges coming down from the middle?

A volcano erupts! Clouds of steam and other gases burst out of the hole at the top. Red hot molten rock comes pouring out. Some of it is blown thousands of feet into the air, to fall miles away as dust and ashes. The thunderous roar of the explosion can be heard hundreds of miles away.

Suppose you had been at school in ancient Greece two or three thousand years ago. How would you have explained the volcano?

You would probably have said that the proper place for fire is high above the sky where the sun is – that is what people believed. It would have seemed obvious to you that some fire had been trapped underground – perhaps the gods had been up to mischief! Now the fire was bursting out to get back to its right place.

The explanation may sound silly nowadays, but it was a good theory for its time. It gave a reason for everything that people could see. If our theories are different, it's because we have seen much more. We have many more observations to explain. Later, you will see how we explain volcanoes today.

Nowadays, we know much more about the different kinds of rocks in the Earth's 'crust'. We can measure movements both in the crust and deep underground, and we can measure the vibrations caused by explosions and earthquakes. This chapter tells about these observations and about the theories scientists have invented to explain them.

2.2 Inside Earth

Back in 1823, an amateur scientist, Captain John Symmes, spoke to the Congress of the United States (something like our House of Commons). He explained his theory that the Earth was made up of five hollow balls nested one inside the other. People lived on each of the balls. There were openings at the North and South Poles so that they could get out on to the surface. Captain Symmes wanted to make an expedition to the North Pole to find the way inside. Twenty-five Members of Congress actually voted to help him.

At the time, there was no evidence to say what the Earth was like inside, so Captain Symmes' theory was as good as any other.

Surprisingly, our modern picture of the inside of the Earth shows it made

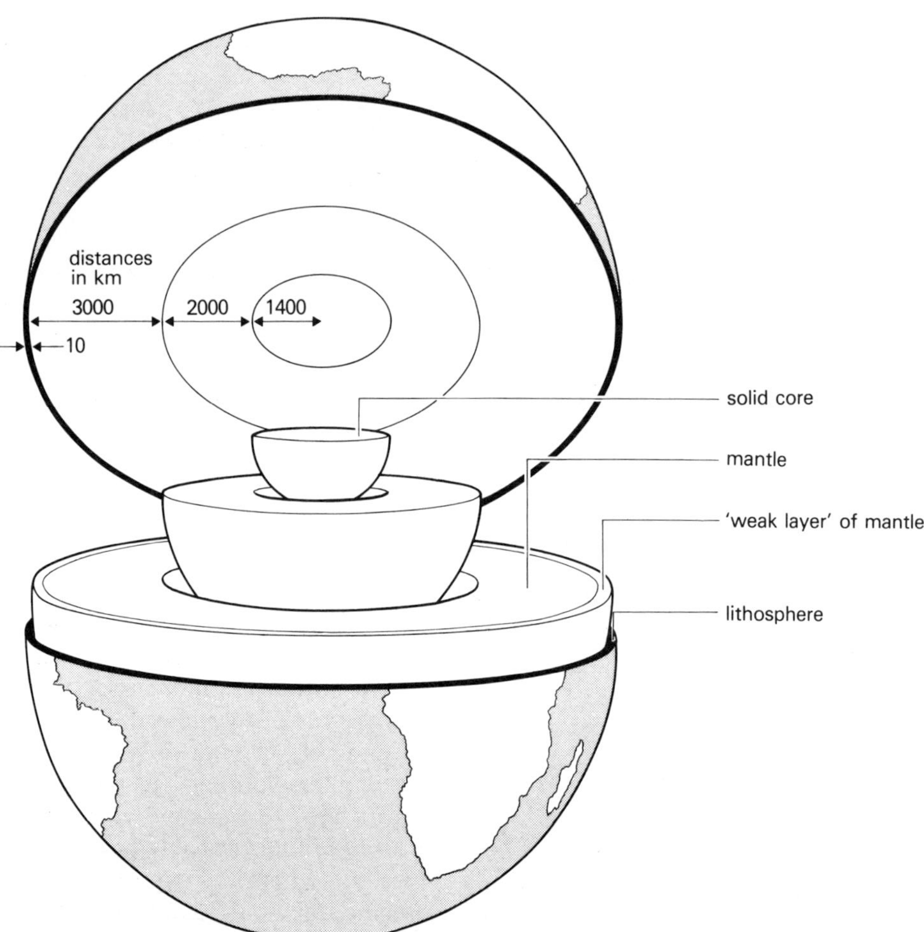

Fig. 2.2 The modern picture of the inside of the Earth

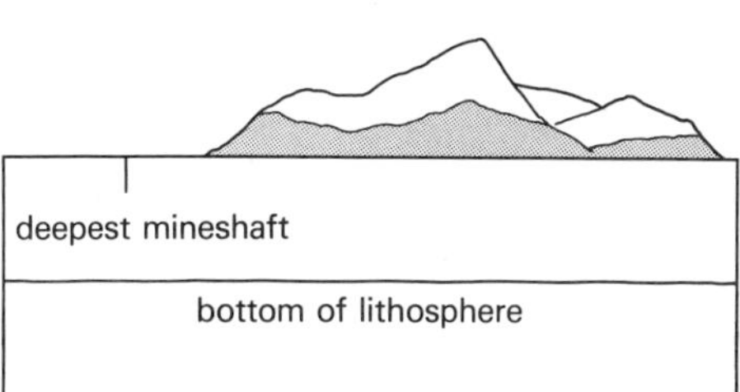

Fig. 2.3 The last 10 cm of a 641 cm strip chart showing the structure of the Earth

Fig. 2.4 The Lisbon earthquake of 1755

up of a set of nesting balls. We do not think that any of the balls are hollow, and we certainly do not believe there are people living inside!

You can get an idea of the thicknesses of the different layers by making a strip chart. You will need a paper tally roll.

1 Measure off a strip of paper 641 cm long. Stick it on a wall horizontally.
2 The left hand end of the strip represents the centre of the Earth. Starting there, measure off 140 cm to the right and draw a line across the strip. That shows the top of the 'core'.
3 Measure 200 cm further and draw another line to show the bottom of the 'mantle'.
4 Draw one more line 300 cm further to stand for the top of the 'mantle'.
5 You should be left with 1 cm for the outermost layer (called the *lithosphere*). You can draw a line 3 mm long to show the deepest mine shaft, and you can stick on a triangle 9 mm high to show Mount Everest.
6 Stand back and look at the strip, which only shows half the Earth. Just think that all the life on Earth goes on in a layer less than a centimetre wide on the chart.

If our deepest mines go down only about 3 km (about 5 miles) and if the deepest oil wells go only a little further, how do scientists know about the mantle and the core? Is it just guesswork? Read on.

2.3 Rock vibrations

The data we have used to build our model of the Earth have come from two main sources, earthquakes and man-made explosions.

In 1755 there was a disastrous earthquake around Lisbon in Portugal. Hundreds of thousands of people died. Shortly after, John Michell, an English earth scientist (*geologist*) began to keep records of vibrations of the Earth. He hoped to find patterns which could be used to predict new earthquakes. He found the work difficult. It was hard to judge the strength of a vibration and hard to say exactly when it happened – especially since you weren't expecting it!

Fig. 2.5 A simplified diagram of a seismograph

Then in 1880, another English geologist, John Milne, invented the *seismograph*. This is an instrument which automatically records the times and strengths of vibrations on a paper strip.

Since the time of John Milne, millions of recordings have been made all over the world. There are over 500 000 earthquakes every year (only a few serious), and man-made explosions give additional readings.

In the last few years, laser beam seismographs have made the readings much more sensitive and much more accurate. These can measure movements of the crust of less than a ten-thousandth of a millimetre.

Whenever there is an earthquake or an explosion, the whole Earth rings like a bell. Vibrations travel in all directions. Some go deep into the Earth, bounce off the different layers and come back up again.

A typical trace from a seismograph can be seen in Fig. 2.6. It shows three sets of vibrations picked up from the *same* earthquake. Which vibration travelled fastest? Which vibration came across the surface? Can you say?

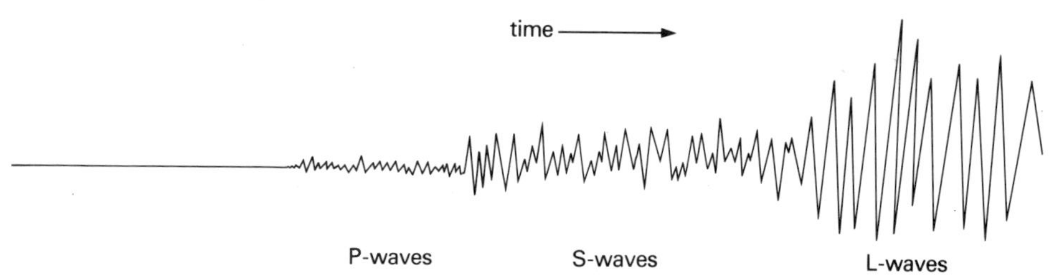

Fig. 2.6 Typical seismograph trace

IN YOUR NOTEBOOK

✳ Write a heading: *Recording earthquakes*

✳ Draw a diagram of a seismograph. Label it.

✳ Copy the seismograph trace.

✳ Write these statements, filling in the gaps:

A seismograph records ______ sent out from the centre of an earthquake. Looking at the trace, I think vibration ______ travelled fastest. I think vibration ______ came across the surface.

From seismograph readings taken at different stations, scientists can work out what the Earth is like inside.

2.4 Minerals

You have often heard the word *mineral*. To most people, it means 'any material which comes out of the Earth'. 'Mineral water' is supposedly health-giving water from natural springs; coal is often called a 'valuable mineral'.

The word has a special meaning for an earth scientist. A mineral is a natural material which has a *definite chemical composition*. That is, it is always made up of the same chemical elements in the same proportions. It is often found as crystals.

Rocks are made of minerals. The colours you see in granite are probably caused by different minerals.

The following table lists some important minerals and gives their chemical names.

Mineral	Chemical name	Uses
corundum	aluminium oxide	
copper	copper	
galena	lead sulphide	
gold	gold	
gypsum	calcium sulphate	
hematite	iron oxide	
halite	sodium chloride	
malachite	copper carbonate	
quartz	silicon dioxide	
sulphur	sulphur	

IN YOUR NOTEBOOK

* Write a heading: *Minerals*

* Copy out the table. Try to fill in the gaps in the third column. Some research in the library will help.

* Under the table, complete this statement:

    To a scientist, the word mineral means . . .

Most minerals are useful, so most of them are valuable. Searching for them has caused wars, fighting and great journeys of exploration. Some of them have made great changes in our lives. For instance, you will see how important sulphur is to modern life and industry on page 110.

Geologists have designed tests to sort out the minerals. Some of the tests are shown in the next section.

Investigation 2.1 Tests for minerals

You will need a small collection of minerals such as: felspar, calcite, galena, malachite, mica, hematite, and talc.

IN YOUR NOTEBOOK

* Make a table like the one overleaf for *each* of the minerals you test:

Mineral

1 Colour	2 Streak	3 Lustre	4 Cleavage	5 Hardness

Colour test

Relying on colour is not a good way of identifying minerals. Their colours are too easily changed by tiny impurities of other chemicals. Still, it might suggest a 'range of possibilities'.

Streak test

Rub the mineral across a piece of white, unglazed, pottery. Write down the colour of the powder mark (the 'streak') if the mineral is soft enough to make one.

Lustre test

Turn the mineral back and forward in good light near a window (not in direct sunlight). Decide how you would describe the way it shines. Is it *glassy, greasy* (like candle wax), *metallic, resinous* (like toffee) or just plain *dull*? Choose one of the words and write it down.

Cleavage test

Tap the mineral *gently* with a hammer and a single-edged razor blade. *Wear safety specs, point the blade away from yourself and take care not to break it.* Does the mineral split into flat-sided flakes or regular crystal shapes? Try tapping it in another direction. If it does 'cleave', make a sketch of the shape which appears.

Hardness test

Some minerals are harder than others. You can test them by trying to scratch them with various objects. The result gives you a number for the hardness of the mineral.

Scratched by:	Hardness
Finger nail	1
Cupro-nickel coin (10p)	2
Knife blade (easily)	3
Knife blade (with difficulty)	4
File (easily)	5
File (with difficulty)	6

Minerals harder than level six will scratch window glass. *Do not try them out on the school windows!*

2.5 Birth of the rocks

How were rocks and minerals formed? Where did they come from? Are they still being formed? Why are there so many different kinds? Why are they different?

We can begin to find answers to questions like this by taking another look

Fig. 2.7 Volcanic formation on Bali, Indonesia. This photo was taken from the air using RADAR.

at volcanoes. Fig. 2.7 shows one on the island of Bali. You can clearly see the tracks made by old flows of lava.

Since the 1960's, scientists have worked out a theory which says that the outermost layer of the Earth is made up of great plates floating on very hot rock beneath.

Some of the edges of the great plates grind against each other all the time. The stretching, colliding and sliding cause earthquakes and cracking of the rocks. The great pressures in the lithosphere force molten rock (called *magma*) up through weak spots. Any water the magma meets is turned instantly into steam – increasing the pressure even more. Sometimes the molten rock and steam reach the surface and a volcano erupts.

If you could slice an active volcano in half you would see something like Fig. 2.8. The magma cools as it is forced up. When it cools below about 1000 °C it solidifies. If the cooling is slow, crystals form in the rock – the slower the cooling, the larger the crystals. If the cooling is sudden, crystals do not have time to form.

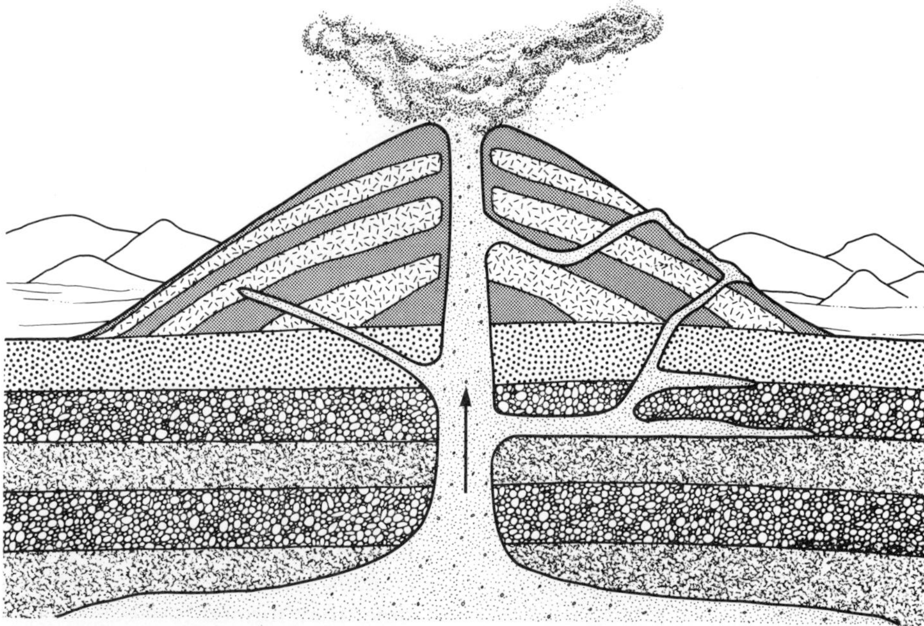

Fig. 2.8 Section through an active volcano

Look at Fig. 2.8 again. Can you point to a place where the magma is cooling very quickly indeed? Can you point to a place where the magma is trapped and cooling slowly? When you have decided, read the answers on page 161. Then do the work in your notebook . . .

IN YOUR NOTEBOOK

✱ Write a heading: *The birth of the rocks*

✱ Draw a diagram of a volcano. Put these labels on your diagram:

molten rock (magma);
lava;
obsidian (glassy rock, no visible crystals) being formed here;
granite (with large crystals) being formed here.

✱ Write this important statement:

Rocks such as granite, obsidian, and basalt are formed when hot molten rock comes up through the Earth's crust and freezes. These rocks are called igneous rocks.

All the rocks in the crust of today, started out as magma. Over the years, some of them were slowly worn down into tiny bits. Can you say how? There are clues in Fig. 2.9.

Fig. 2.9 Niagara Falls. What effect is the water having on the surrounding rocks? (Look at the sky too.)

Wind and rain, rushing water, ice, the frosts of winter and the heat of summer all played their part in cutting away and grinding down the rocks. The process is called *erosion*. The result was pebbles, sand, clay, and mud.

Water and ice carried the materials down into the hollows of the Earth – under the seas for example. As they settled on the bottom they were pressed and cemented into new rocks. Sometimes the bodies of plants and animals settled at the same time, and were surrounded by the rocks. In time, they formed fossils.

Quantities of small particles which settle out in layers are called *sediments*.

Fig. 2.10

These new rocks are called *sedimentary rocks*. You can see the pebbles cemented together in rock (a). Rock (b) still shows signs of water ripples. Rock (c) has a fossil in it.

(a) Sedimentary rock made up from pebble sized rocks

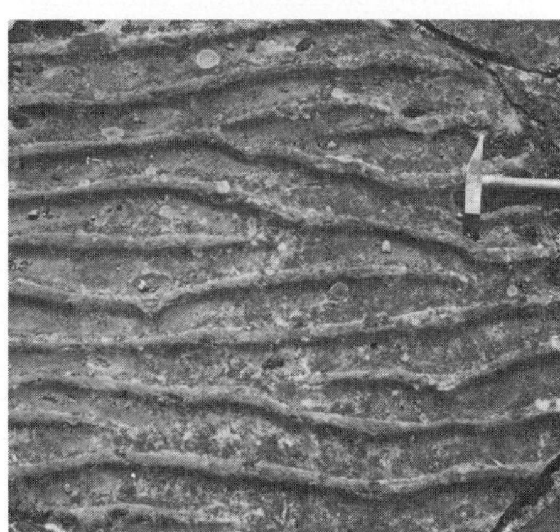

(b) Sedimentary rock with signs of water ripples

(c) Sedimentary rock with fossil

IN YOUR NOTEBOOK

✳ Write these statements, filling in the gaps:

Rocks such as limestone, sandstone, and conglomerate are _______ rocks. They were formed when other rocks were eroded, carried along by _______ and _______ and laid down in hollows. Their particles were pressed and cemented together. Sometimes parts of the bodies of plants and animals were mixed in. These formed f _______ .

2.6 Changed rocks

You can see that these cliffs were formed from layers of sediment. The layers are called *strata*. Which are the oldest strata? Which were laid down first? Read the answer on page 161 when you have decided.

Fig. 2.11 Sedimentary rocks in a cliff face

IN YOUR NOTEBOOK

✳ Draw a sketch of the cliffs in Fig. 2.11. Label the oldest rocks and the youngest rocks.

The life story of the rocks did not end when the sedimentary rocks were formed. It is still going on today. The plates of the Earth's crust have carried on moving, grinding and pushing against each other. Magma has carried on pressing from underneath. The sedimentary layers have been squeezed, twisted, bent and broken. Look at Fig. 2.12.

In Fig. 2.12(a) the rocks have been folded. Notice that some of the surface rocks have moved down and new sediments have been laid on top of them.

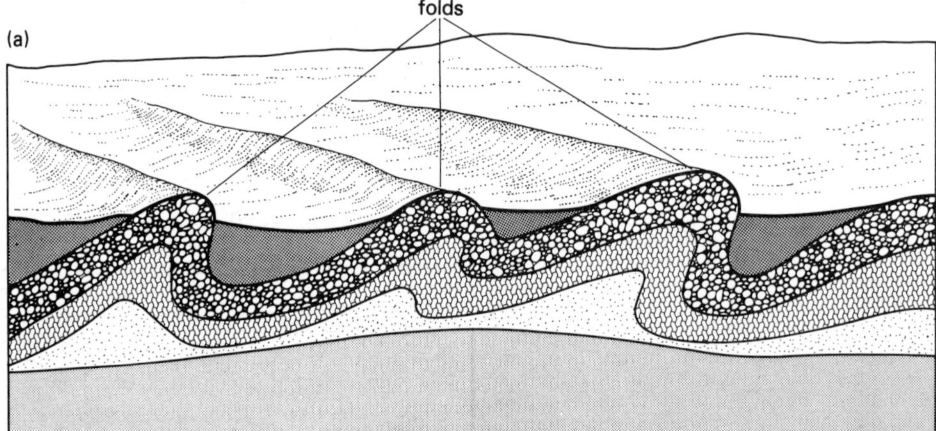

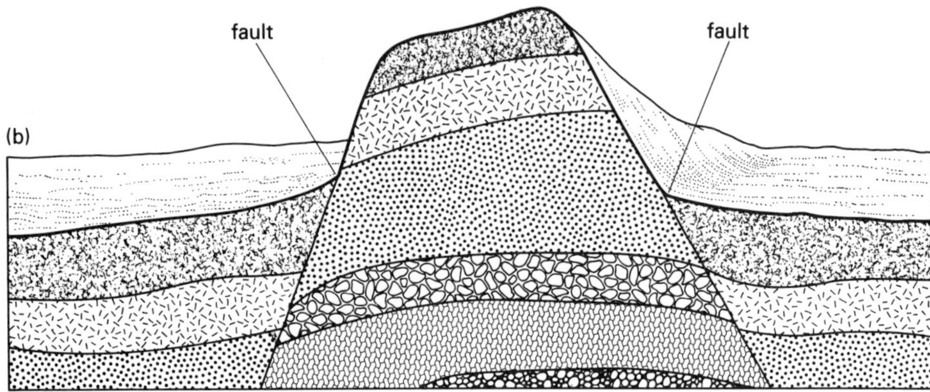

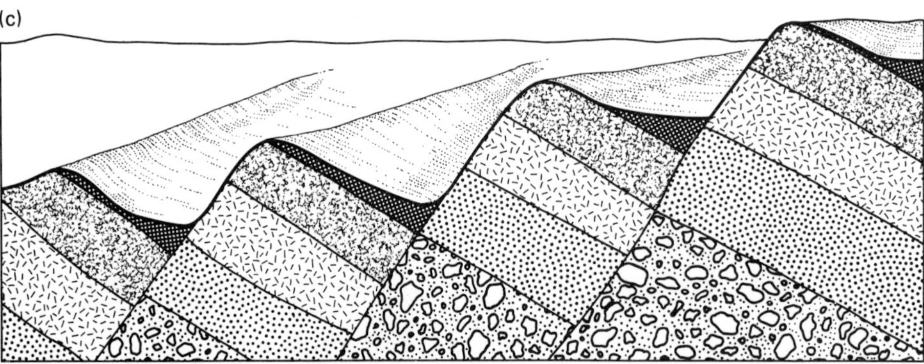

Fig. 2.12 (a) Folded rocks
(b) Two faults with a mountain pushed up between them
(c) A mountain range formed along many faults

In Fig. 2.12(b) two cracks called *faults* have formed and a mountain has been pushed up.

In Fig. 2.12(c) many faults have led to the forming of a mountain range.

You can imagine how folding, faulting, and the weight of new sediments might press down some of the older sedimentary rocks (the ones near the bottom). This has happened. As the sedimentary rocks were buried they were put under enormous pressure and heated until they softened or melted. Chemical reactions took place in some of them. Some mixed with the magma.

Later, the rocks were lifted again – by more folding and faulting. The rocks looked very different. Such *changed* rocks are called *metamorphic* rocks. A particular kind of metamorphic rock can be made from several different kinds of sedimentary or igneous rocks. The rock *gneiss* can be changed basalt or granite or sandstone or shale. The rock *marble* can be changed chalk or limestone.

Fig. 2.13

(a) Granite

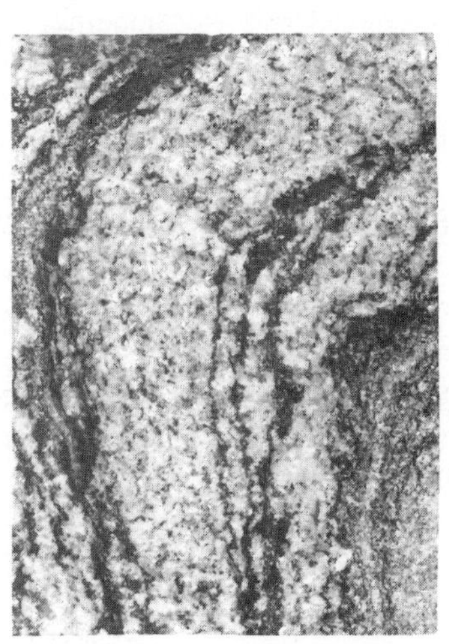

(b) Gneiss

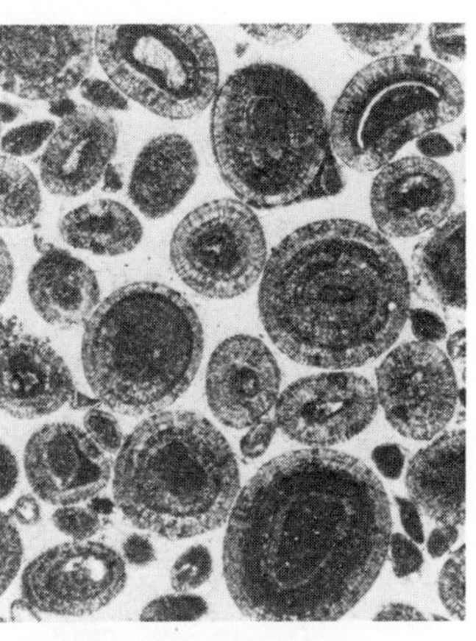

(c) Limestone (a photomicrograph of oolitic limestone, at × 25 magnification)

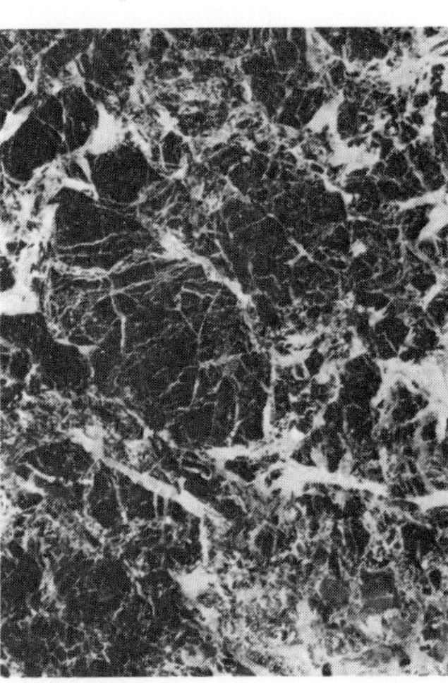

(d) Marble

IN YOUR NOTEBOOK

✳ Write these statements, filling in the gaps:

Rocks such as schist, marble, and slate are formed when igneous or sedimentary rocks are heated and put under great _______. They are called m _______ rocks.

✳ Copy this table and fill in the gaps. You may need to write the names of several rocks in each. (Some library research may be needed.)

Original rock	Metamorphic rock
granite, basalt, sandstone, or shale	gneiss
_________	quartzite
	slate
limestone or chalk	_________
basalt	schist or g _________

Sedimentary rocks which become melted and mixed with the magma will form new igneous rocks. Perhaps they will be eroded again and form sedimentary rocks. Then they may be pushed down to form metamorphic rocks – or perhaps mix with the magma . . . and so on, round and round.

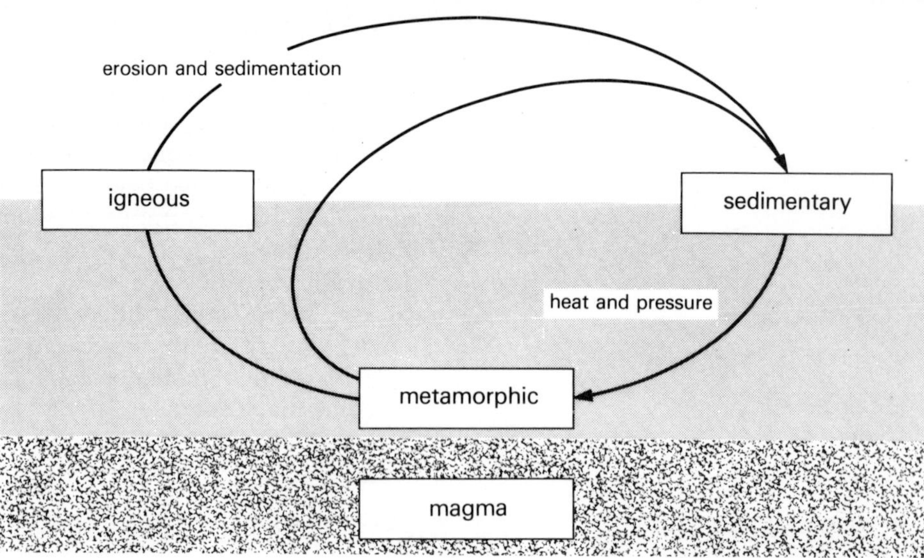

Fig. 2.14 The rock cycle

The life story of the rocks does not have a clear beginning and end. It goes on and on, round and round. Nowadays, the *rock cycle* is turning over more slowly than it did a few million years ago but it has not stopped – earthquakes remind us of that.

It is almost impossible to imagine the vast stretch of time which has passed since the rock cycle began turning. One way to help our imaginations is to use a little science fiction. Read on . . .

2.7 Time squeeze

Imagine that TV cameras had been around when the Earth was forming (about $4\frac{1}{2}$ thousand million years ago), recording everything on tape. Perhaps they were in an 'alien' space craft! Imagine that somebody decided to play back all the tapes in a year stopping them once a day to see what was happening on Earth. Of course, that would mean running the tapes $4\frac{1}{2}$ thousand million times faster than normal – but perhaps the aliens could do that. They would see the first signs of life in April of their year, and humans like us would not appear on the tape until 10 minutes to midnight on 31 December!

2.8 The most important minerals

If you completed the table on page 33 you will know that the minerals: galena, hematite, and malachite are particularly important to us. Can you say why? What do they have in common?

The answer is that we can extract metals from them – lead from galena, iron from hematite, and copper from malachite. Such minerals are called *ores*.

In some parts of the world it is possible – if you look hard enough – to pick up bits of almost pure copper, silver, and gold. (The word 'metal' comes from a Greek word meaning 'to search'.) Six thousand years ago those were almost the only metals that people had. They did not know how to get the metals from their ores. Some very important people – rulers and priests, for instance – had small and highly valued ornaments made from iron. Iron is never found like silver and gold, so can you guess where it came from?

The iron had fallen from the sky as meteorites! People thought it was a present from the gods!

Then suddenly, about 5000 years ago, an important discovery was made. Perhaps somewhere around the part of the world we call Turkey, people were raking over the ashes of a fire. They spotted some tiny, bright beads of copper. They realised that the copper had come from the rocks used to make the fireplace. Heating the rocks with charcoal in the fire had set free the metal from its ore. Such a process is called *smelting*. You can investigate this in the lab.

Investigation 2.2 Smelting

You will need:

a spatula	small beaker of water
a tin lid	crushed malachite
Bunsen burner, tripod, and gauze	charcoal

1 Put a spatula of malachite on the tin lid. Heat it strongly for about five minutes.
2 Let the lid cool for a few minutes. Tip the roasted ore into the beaker of water. Look for signs of copper metal on the lid and in the beaker.
3 Mix one spatula of malachite with $1\frac{1}{2}$ spatulas of charcoal.
4 Put one spatula of the mixture on the tin lid. Cover it with more charcoal. Heat it strongly for about five minutes.
5 Again, tip the product into a fresh beaker of water and look for signs of copper.

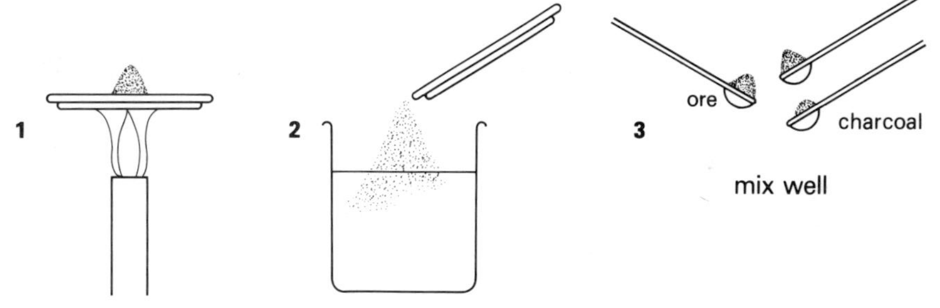

Fig. 2.15 Investigation 2.2

IN YOUR NOTEBOOK

✳ Write a heading: *Smelting copper ore*

✳ Draw a diagram of the investigation.

✳ Write these statements, filling in the gaps:

When I heated copper ore on its own, I could see _______ traces of copper at the end.
When I heated the ore with charcoal, I could see _______ at the end.
Heating a metal ore with charcoal is called s _______.

You can try smelting galena to get lead metal. Heat the crushed galena on its own for about five minutes. Let it cool, mix it with charcoal and heat it again. Look for drops of melted lead.

Scientists explain smelting by saying that the ores contain *compounds* of the metals. That means that the metals are joined up to other elements such as the gas oxygen. When the ores are heated with charcoal (*carbon*) the

Fig. 2.17 Flint tools and weapons made by early people

Fig. 2.18 Bronze weapons made by bronze age people

elements 'change partners', leaving the metal behind (look at Fig. 2.16). You will read more about this in Chapter 6.

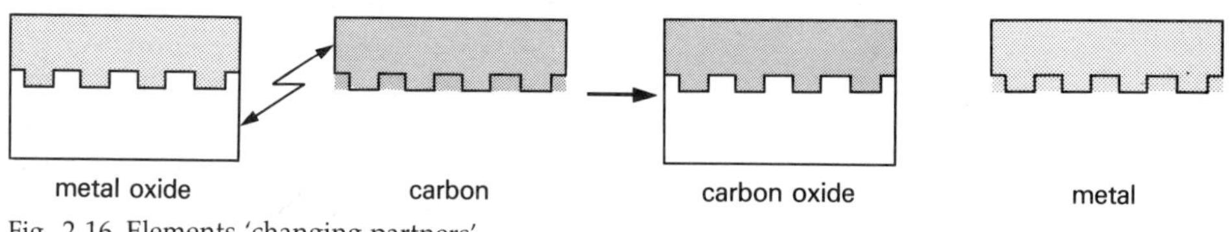

Fig. 2.16 Elements 'changing partners'

2.9 Metals in history

People similar to us have been living on Earth for about 50 000 years. It's easy to talk about 50 000 years, but very difficult to imagine what it *means*!

It helps our understanding if we divide such long time periods into 'lifetimes' of about sixty years each. So we can say that people like us have been around for about 800 lifetimes. For most of that time, people lived in caves. They started to build houses about 150 lifetimes ago.

As the lifetimes went by, people learned to make beautiful tools and weapons from rocks and minerals. Flint was the most popular mineral. Then about 80 lifetimes ago they learned to get copper from the rocks.

Another eight lifetimes had to pass before people in the East discovered *alloys* – mixtures of metals. They found that if they mixed copper ore and tin ore and smelted them together they got a new metal. It was stronger and harder than copper. And it was easier to make into shapes because it could be poured into a mould ('cast'). We call it *bronze*.

The 'Bronze Age' lasted for about 30 lifetimes. The marvellous new tools and weapons gradually found their way into Western Europe.

People began trying to smelt iron ore during the Bronze Age (about 60 lifetimes ago). It did not work very well at first. Their furnaces were not hot enough to melt iron. All they got was a half-melted material looking like a hard black sponge. Still, as more lifetimes passed, they learned to work this into usable iron. The trick was to hammer the material as it cooled, put it back into the furnace with more charcoal, take it out and hammer it again, and so on. Thirty lifetimes ago, Roman soldiers were marching into Britain, iron swords at the ready.

The smiths gradually learned to make their furnaces taller and to blow air in with bellows. About ten lifetimes ago they started to use waterwheels to operate them. Soon after, they produced the first *molten* iron – which they could cast in a whole range of new shapes. By the time King Charles I was born (about 6 lifetimes ago) iron was cheap and plentiful.

By the time King Charles II had died, a hundred years later, the position had changed. Most of the great forests of Southern England had been cut down, and wood charcoal was becoming very expensive. The problem was solved by an 'iron master' called Abraham Darby at his works at Coalbrookdale in Shropshire. He loaded his iron furnaces with a new fuel – *coal*. You can still see the cast iron bridge he built to mark the achievement.

Investigation 2.3 A metals strip-chart

You can show the 'milestones' in the history of metals on a strip chart. Make it from a tally roll like the one on page 31, but remember that this one is showing time periods, not distances.

1 Measure off a length of tally roll paper 800 cm long. The scale is 1 cm = 1 human lifetime. Fasten it up on a wall.
2 Measure off 650 cm from the *left* end of the strip and draw a line. Label the line: *People stop living in caves*.
3 Measure off 80 cm from the *right* hand end and draw a line. That stands for our guess at the time, eighty lifetimes ago, when people learned how to smelt metal ores. Label the line: *Smelting discovered*.
4 Put these other 'milestones' on the strip:

'Milestone'	Lifetimes ago
First alloy-bronze	70
End of bronze age	40
First tries at iron smelting	60
Waterwheel-driven bellows	$10\frac{1}{2}$
Cast iron being made	9
Coke instead of charcoal in furnaces	4
First iron ship	$2\frac{1}{2}$
First army tanks	1

IN YOUR NOTEBOOK

✳ Write a heading: *Metals in history*

✳ Write a letter to an imaginary friend describing how you made the metals time strip. Tell him or her what the strip looked like when you had finished it. Write about anything that particularly surprised or interested you.

2.10 Blast furnace

Fig. 2.19 Port Talbot in South Wales

This is Port Talbot in South Wales. The town is famous as an iron-making centre. The towers in the background are modern iron furnaces – called *blast furnaces*. Inside, they look like Fig. 2.20.

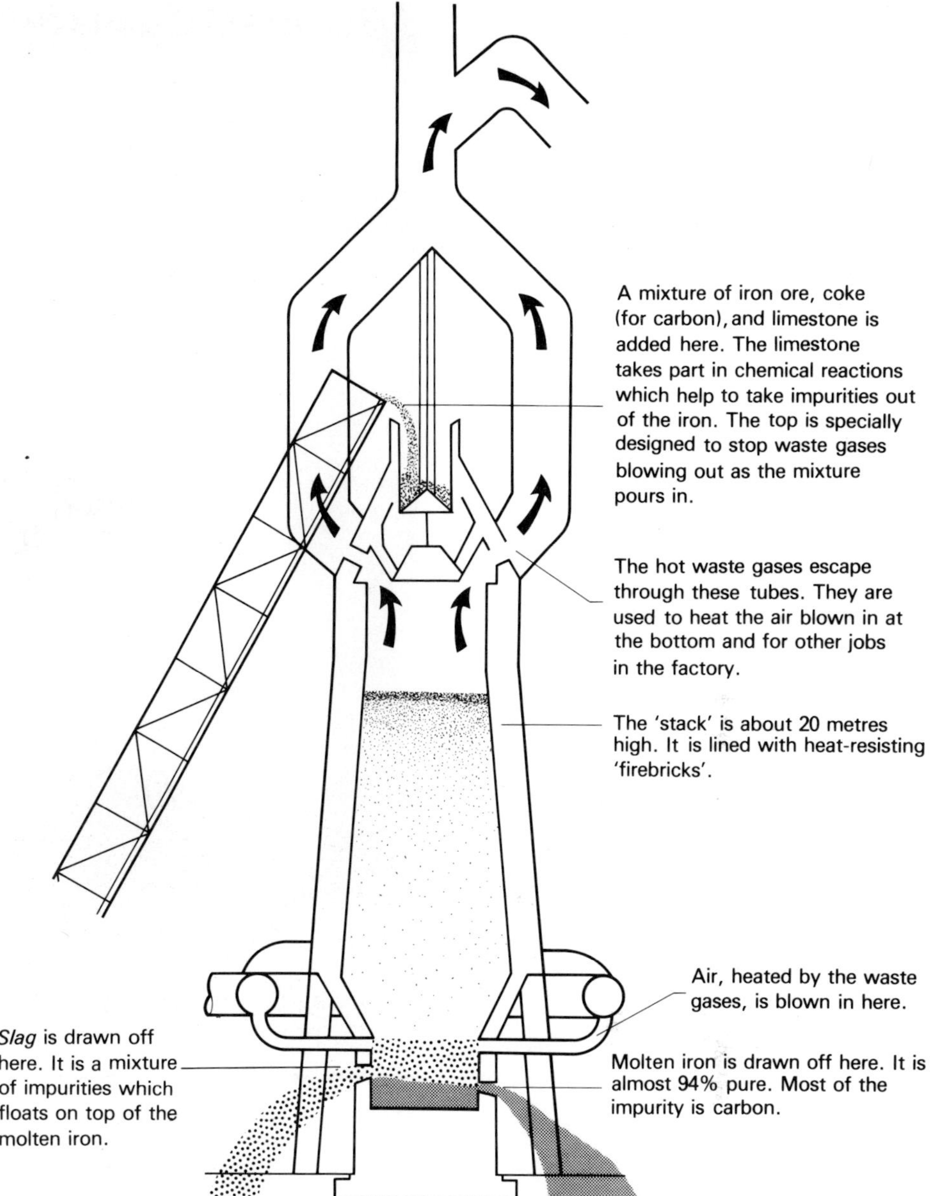

Fig. 2.20 A blast furnace

The problem with the iron from the blast furnace is that it is brittle. It breaks easily if it is bent or twisted. It can be cast to make things like manhole covers and cooking pots, but is no use for car bodies and kitchen knives. It is brittle mainly because it contains too many impurities such as carbon.

In 1856 an English metal scientist (*metallurgist*) worked out a method of burning out the impurities in the blast furnace iron. His idea was to blow more air through the molten iron after it had been run out of the furnace. The 'experts' said he was mad – the cold air would simply freeze the iron!

When he tried it, the result was surprising and a little frightening! The carbon *did* burn in the air. Not only that, the heat from the reaction raised the temperature even more. The result was something like an explosion. When it had died down he was left with iron with very little carbon in it. Such iron is called *steel*.

The metallurgist's name was Henry Bessemer (later, *Sir* Henry – he was knighted for his achievement). Fig. 2.21 shows hot iron being poured into an oxygen furnace which uses his design.

Fig. 2.21 Pouring hot metal into a steelmaking vessel at a steelworks

Varying the amount of carbon in steel, and adding small amounts of other metals, produces steels with different properties – suitable for different jobs.

IN YOUR NOTEBOOK

✳ Write a heading: *Making iron in a blast furnace*

✳ Draw a diagram of a blast furnace. Label the different parts.

✳ Write another heading: *Uses of steel*

✳ Write down ten different things made of iron or steel in and around your home. Do not forget the garage and garden. When you have finished, compare your list with your friends' lists. Have your friends spotted things which you missed?

✳ Look in newspapers and magazines to find pictures which show 'steel in use'. Stick the picture into your notebook.

2.11 When will the Earth run out?

Metal ores are minerals. Minerals come out of the Earth. Every day we dig more of them out. Our mines get bigger and deeper. How long will it be before they run out? How many lifetimes before we have used up all our metal ores?

Fig. 2.22 Bougainville copper working, Papua New Guinea

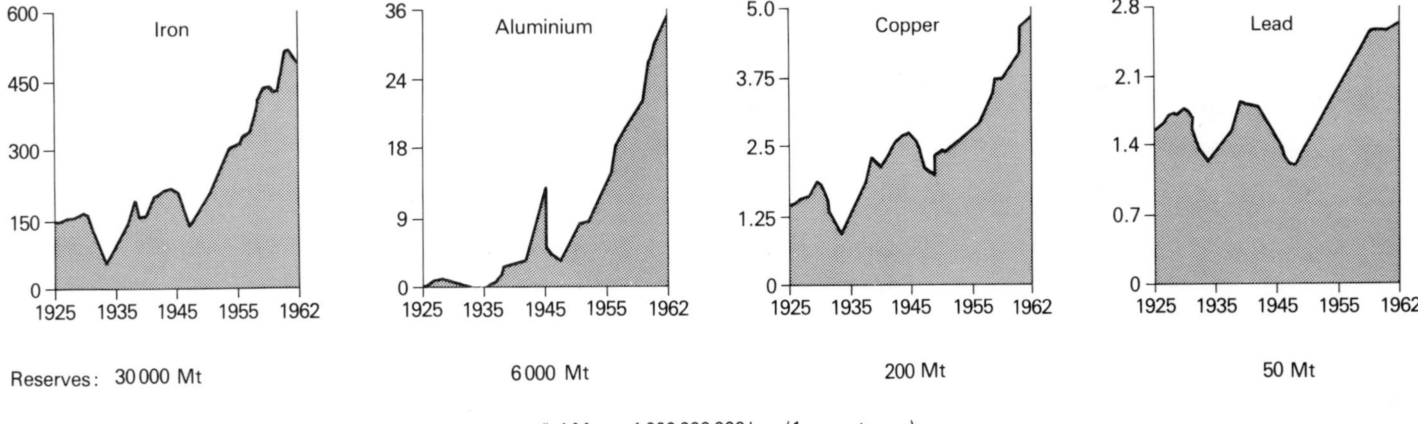

Fig. 2.23 World production of iron, aluminium, copper and lead

In 1962, an international survey was carried out to try to find the answer to these questions. The graphs show some of the results of the survey. You can see that in 1962, all over the world, we dug out about 500 million tonnes of iron ore (500 megatonnes, Mt). Under the graph, you can see that the estimated world 'reserves' – the amount of ore left – was 30 000 Mt. If we had kept on digging out iron ore at the 1962 rate, how many years supply would we have had? Can you work it out? (30 000 Mt at 500 Mt a year.)

The answer is that in 1962 we looked to have about sixty years supplies of iron – enough for one lifetime! Things were worse with copper. We were using less but we had less to dig out. The estimate came to a forty year supply.

Since 1962 there has been good news and bad news. The good news is that some reserves have turned out to be greater than we thought. The bad news is that our rate of using the reserves has increased.

Copper, our oldest metal, has become very scarce and very expensive. So much so that aluminium has had to take over from it as the metal for electric wiring. You will read more about aluminium production later.

Nowadays, there are two sorts of answers to the question: When will the Earth run out?

Optimists say that it will be *hundreds* or *thousands* of years. After all, they say, the deepest mine only goes down 3 km and the Earth's crust is 10 km thick. And our prospectors are sure to find new supplies in unexplored parts of the Earth. In any case, we have woken up to the shortage – haven't we? We will use less in the future anyway.

Pessimists point out that water covers two-thirds of the Earth's crust and we have already used up most of the *easy* supplies in the other third.

They say that exploring, finding new places to mine, and digging deeper, are enormously costly. And it uses enormous amounts of energy which we can't spare – coal and oil are running out too! On top of that there's no sign that we *have* woken up to the shortage. Every day we use more and more of our materials.

Fig. 2.24 is a *pessimist's* chart to show how long things will last – the *worst* estimates.

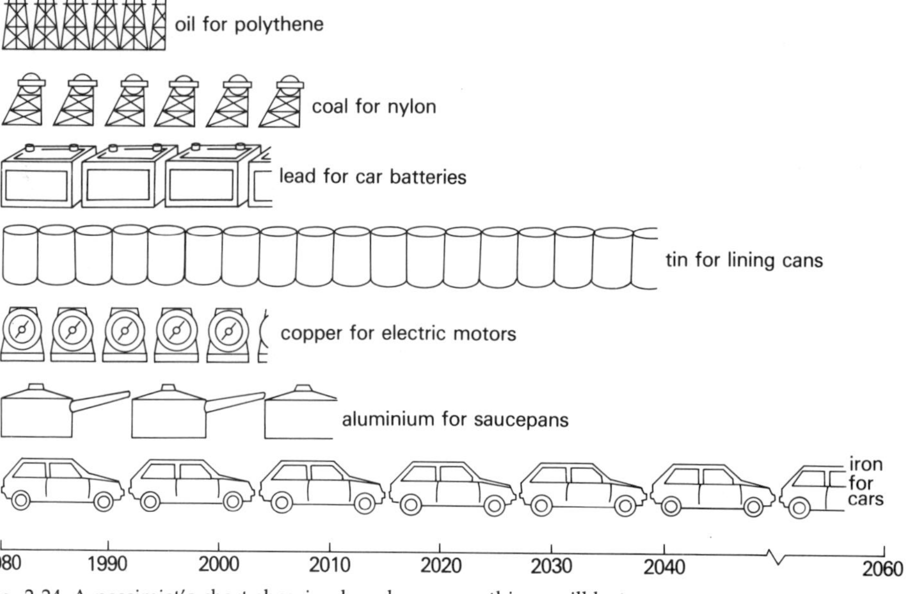

Fig. 2.24 A pessimist's chart showing how long some things will last

IN YOUR NOTEBOOK

* Write a heading: *When will the Earth run out?*

* Copy the 'pessimist's chart'.

* Mark these things on the time scale at the bottom:

  1 The year you will leave school. Draw an arrow pointing to the scale and label it: 'I leave school'.
  2 The year your children start school. If you have children one day, you can estimate that they will go to school when you are about thirty.
  3 The year your children leave school. You can estimate that they will go to school for about eleven years (counting primary school).
  4 The year you retire. Probably when you are about 60.

* Write this statement under the chart, filling in the gap.

   The chart shows coal and oil being used to make plastics (polythene and nylon for example). They are also used as sources of ______.

* How might science be able to help in solving the problem of the minerals shortage? Write down what you think.

Checkout

Keywords

alloy
chemical composition
cleavage
compound
crystal
earthquake
Earth's crust
element
erosion
fault
fossil
igneous rock
lava
magma
metamorphic rock
mineral
ore
reserves
rock cycle
sedimentary rock
seismograph
smelting
strata
vibration
volcano

Patterns

1 There are patterns in the recordings of vibrations from earthquakes and man-made explosions. These give information about the inner structure of the Earth.
2 Certain natural materials called minerals have the same chemical composition wherever they are found in the Earth's crust.
3 Minerals vary in their colour, lustre, cleavage properties, hardness, and chemical reactions.
4 Rocks can be divided into three types on the basis of their appearance, properties, and the rock formations in which they are found. (Igneous, sedimentary, and metamorphic types.)
5 Some sedimentary rocks contain the remains of ancient organisms (fossils).
6 In undisturbed rocks, the older fossils are found under the younger ones.
7 Some metals can be extracted from their ores by heating them with carbon (smelting).

A theory

(a) Igneous rocks are formed when molten rock (magma) is forced upwards, cools and freezes.
(b) The size of the crystals in an igneous rock shows how slowly it cooled.
(c) Older rocks are eroded into small bits which are carried along by wind and water, and laid down as sediments. The sediments are pressed and cemented together to form new rocks.
(d) Various movements of the crust can subject igneous and sedimentary rocks to heat, pressure, and chemical action. They become changed (metamorphic) rocks.

All these happenings are part of a continuing process: the rock cycle.

True or false?

1 If the Earth was shrunk to the size of an orange the crust would be about as thick as the orange peel.
2 If the temperature under the ground goes up by 1 °C every 10 metres the temperature at the bottom of the deepest mineshaft would be about 300 °C.
3 There is at least one earthquake per minute somewhere in the world.
4 All kinds of granite have the same size crystals.
5 Obsidian is a rock which cooled very slowly as it formed.
6 All the igneous rocks formed before the sedimentary rocks.
7 Fossils are never found in igneous rocks.
8 Another name for granite is gneiss.
9 Bronze is a mixture of copper and iron.
10 If we go on using metals at our present rate and no new reserves are found, most of the common ores will run out within two lifetimes.

Problems

1 *Read the passage and try to answer the questions.*
One of the tourist attractions near Naples in Italy is an extinct volcano. The large crater in the top is filled with water making a lake more than half a mile across. Just above the water level, a cave runs into the hillside which used to be the side of the crater. The cave is the 'Grotto of the Dog', which has been famous for hundreds of years. Inside the grotto, there is no obvious smell and the air is clear except for a thin mist covering the floor, but the whole place has a feeling of mystery.

The grotto earned its reputation, and its name, because of the strange things which happened to dogs and other small animals which found their way into it. Humans could walk about without any ill effects, but small dogs soon showed signs of discomfort and fainted. If they were not rescued they quickly died. There is a story that the cruel Roman Emperor Tiberius once had two slaves chained to the floor as an experiment. They died within minutes – just like the dogs. Can you guess why?

The modern explanation is simply that the floor of the grotto is covered with a thin layer of poisonous gas. You probably guessed.

(a) What does the passage tell you about the poisonous gas? Try to write down three things about it?
(b) What tests would you want to carry out in the cave? What gas do you think is causing the trouble?
(c) How do you account for the gas being there? There is a clue in the passage.

2 *Read this passage from an American magazine 'Science Horizons' and try to answer the questions.*
About 150 million years ago, the Gulf of Mexico was a shallow sea. Water evaporated from it faster than new water flowed in. Gradually, the sea dried up, and salt crystallised from the water to form a vast deposit several thousand feet deep.

To the southwest, in what is now eastern Mexico, the Rio Grande river poured heavy deposits of silt and sand on top of the salt layer. Sand and silt from all the ancient rivers began covering the flat salt plain at rates of up to nine inches a year.

After tens of millions of years, the growing weight of the sand and silt became so great that the ancient Gulf sea floor gave way. It sagged to form a trench with a depth that now averages five miles below sea level.

Fig. 2.25

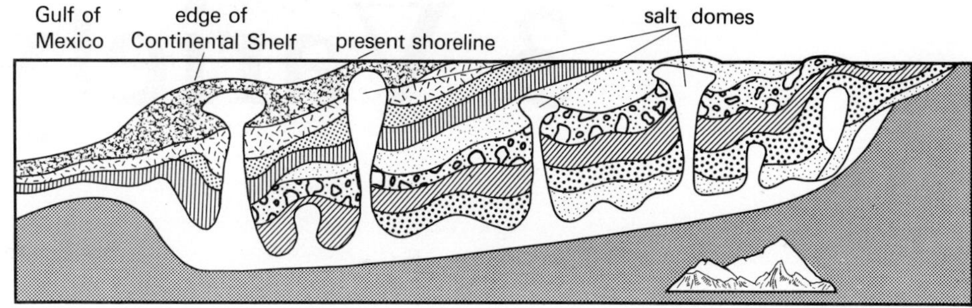

Fig. 2.26 Salt domes in a trench along the Gulf of Mexico

The weight also became unbearable for the salt. It began to melt and flow, like an ocean of thick, hot molasses seeking an escape.

The molten salt forced its way upward through weak spots in the overlying rock layers. The rocks fractured in huge blocks that turned in topsy-turvy patterns. The fractures, faults, and other structures formed traps for oil and gas deposits. The heat also melted sulphur trapped in the rocks, and since it was lighter than rocks or salt, it collected near the upper part of the rising salt masses.

(a) If oil collects in rock formations shaped like upturned letter 'V's, where would you expect to find oil? Sketch part of the diagram to show your answer.
(b) Where would you expect to find sulphur? Show it on your sketch.
(c) Where would you choose to drill a mineshaft to reach the salt? How deep would it be? (The scale is 1 mm = 1½ km). How would the depth of your mine compare with the deepest ones at present?

3 Variety

3.1 All creatures great and small

We live in a world full of *differences* – a world full of *variety*. Often, there seem to be so many differences between things, that it is difficult to spot similarities between them.

Part of the work of scientists is to take away some of the confusion by finding *similarities*. As you saw in Chapter 1, scientists study the world and try to make statements which tell how things are similar. We call these 'similarity statements' *patterns*.

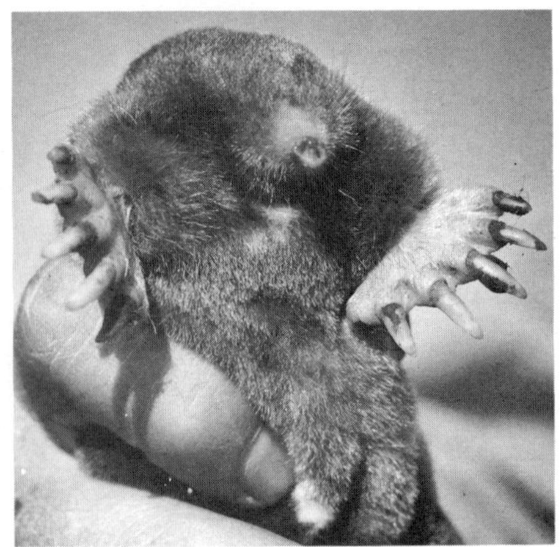

(a)

(b)

(c)

(d)

Fig. 3.1 (a) Common mole
     (b) Star-nosed mole
     (c) Common shrew
     (d) Sloth

For instance, there are patterns which say what different kinds of moles have in common. And there are patterns which say what a mole has in common with a three-toed tree sloth or a shrew.

Then there are much larger patterns. For instance, patterns which answer questions like these:

What do all living things have in common?

Why is there so much variety amongst them?

How does the variety develop?

This chapter is about such larger patterns.

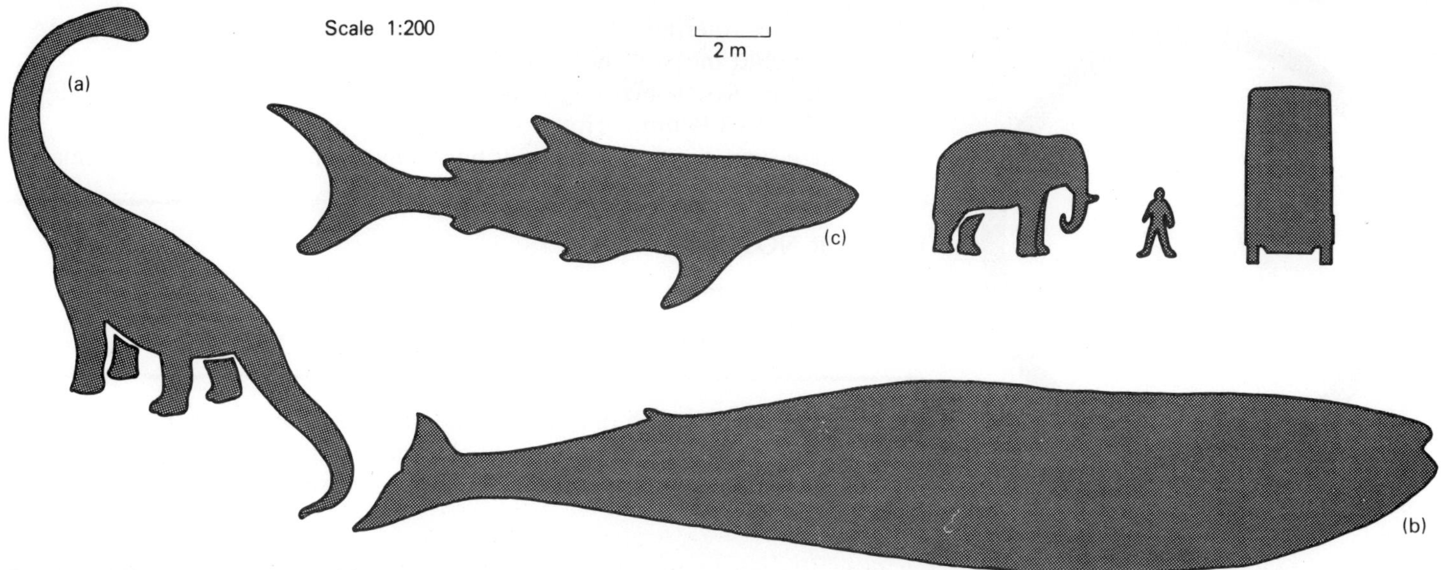

Fig. 3.2 (a) Dinosaur. Disappeared from the earth about 60 million years ago. (Was about 20 metres long.)
(b) Blue whale. The largest animal that ever lived. There are still a few left in our seas. (About 30 metres long.)
(c) Whale shark. The largest living fish. (About 15 metres long.)

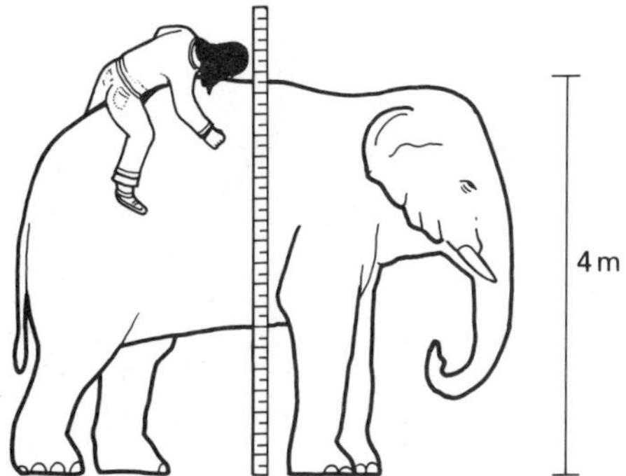

Fig. 3.3 Measuring the height of an elephant!

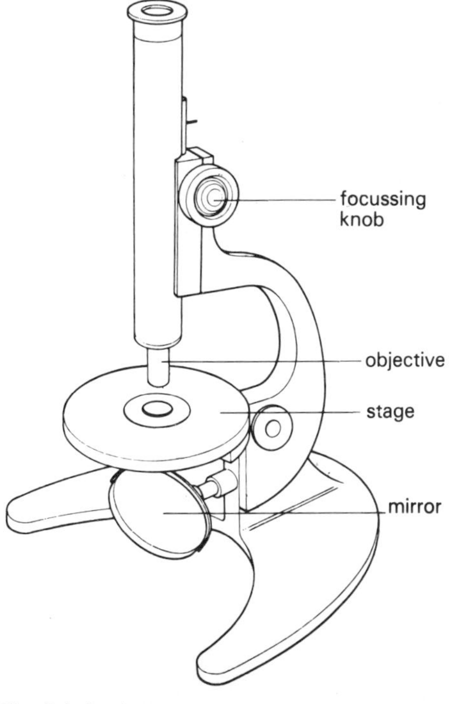

Fig. 3.4 A microscope

One way to get to grips with variety – to try to organise it into patterns – is to look at only one kind of variety at once. For instance, Fig. 3.2 shows some variety in *size*.

You can tell someone how big an animal is by telling them its height or its length. What else might they want to know to do with its 'bigness'? What else could you try to work out?

It is easy enough to measure the height of an elephant (see Fig. 3.3), but how do you measure the height of something very small – a shrimp for instance? You use a special instrument called a microscope. (Fig. 3.4.)

When you look down a microscope, the circle of light that you see is called the *field of view*.

You can find the width of the field of view like this:

* Put a transparent ruler on the microscope stage.
* Look sideways at the microscope. (Do not look through the eyepiece yet.) Wind the tube down until the objective nearly touches the ruler.
* Look through the eyepiece and wind the tube up to focus on the ruler.

The millimetre space on the ruler will look much wider (Fig. 3.5). How many spaces can you see?

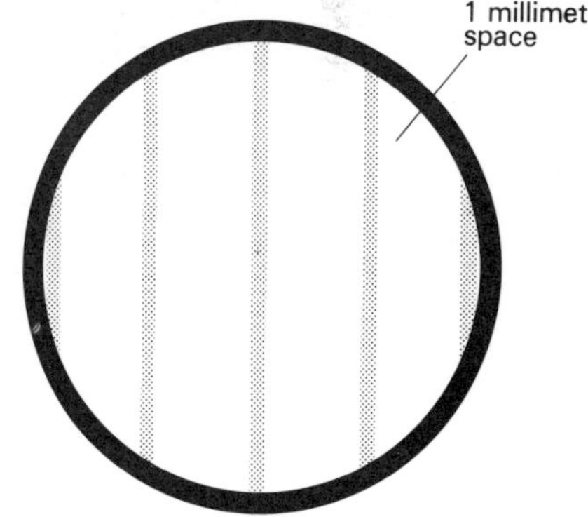

Fig. 3.5 Millimetre marks on a ruler seen down a microscope

IN YOUR NOTEBOOK

* Write a heading: *Estimating the field of view of a microscope*

* Draw a picture of a microscope.

* Write these statements, filling in the gaps:

   Through the microscope I could see _____ *millimetre spaces.*
   The field of view of the microscope is _____ *mm.*

Fig. 3.6 A small animal seen down a microscope

If you took the ruler away and put a very small animal on the microscope stage, you could measure how long the animal is.

Suppose you saw something like Fig. 3.6. And suppose the field of view was 4 millimetres (4 mm). How long would the animal be?

Of course, you cannot say exactly how long. Your answer is an estimate.

IN YOUR NOTEBOOK

✳ Write your estimate:

I estimate the length of the small animal to be _______ *mm.*

Remember that the width of a circle is the same in all directions (look at Fig. 3.7), so you can estimate the animal's width too.

If you use a more powerful objective lens, you can estimate the sizes of even smaller things.

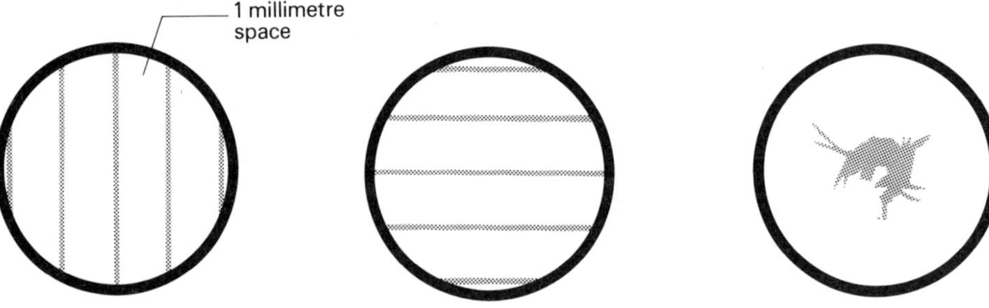

Fig. 3.7 Can you estimate the width of the animal in millimetres?

Investigation 3.1 Looking at some very small things

You will need:
microscope with powerful objective lens
some very small things

Estimate the lengths and widths of the small things you have collected. Make a table of results for your notebook.

3.2 Bony differences 1

Tens of millions of years ago animals like the one in Fig. 3.8 lived on the Earth. Today, the elephant is the largest land animal but whales come much larger. Look at the skeletons of the dinosaur, elephant and whale (Figs 3.8, 3.9 and 3.10).

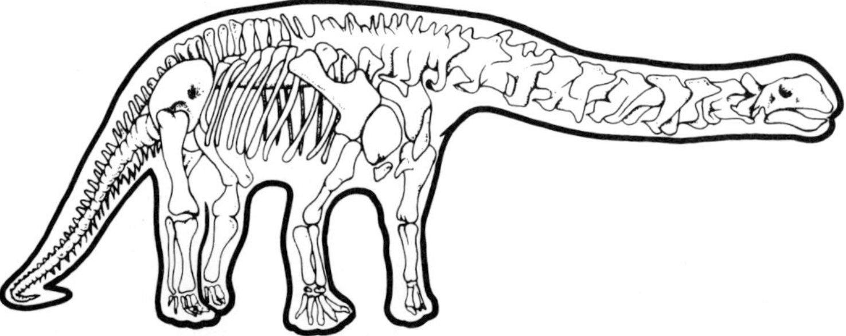

Fig. 3.8 Dinosaur skeleton

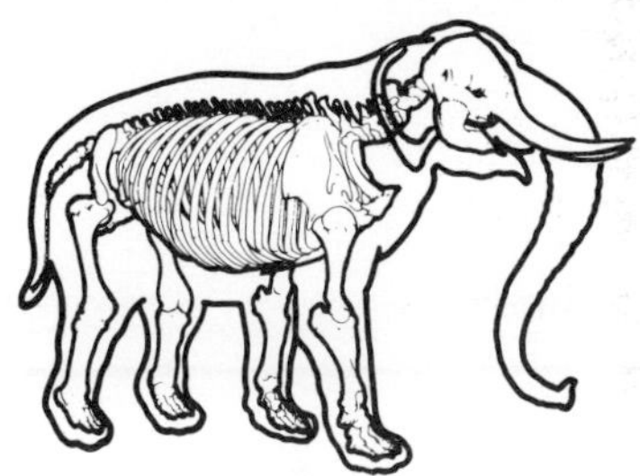

Fig. 3.9 Elephant skeleton

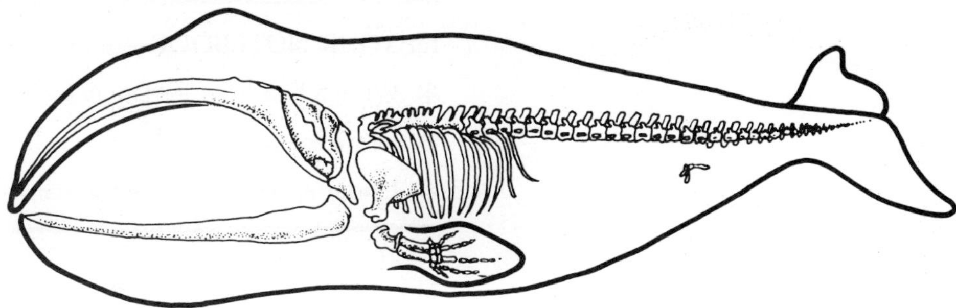

Fig. 3.10 Whale skeleton

What do you notice about the whale's skeleton?

Compared with its overall size, the whale has a much smaller skeleton. Why do you think that is? Try this experiment.

Investigation 3.2 Comparing an 'elephant' with a 'whale'

You will need:
a balloon
a spring balance
bucket of water

1 Weigh a balloon full of water.
2 Look at the balloon's shape 'on land'. (Fig. 3.11.)
3 Weigh the same balloon under water.
4 Look at its shape then.

Can you say why a whale needs a smaller skeleton (for its overall size) than an elephant does?

What is the main danger to a whale if it is stranded on a sand bank?

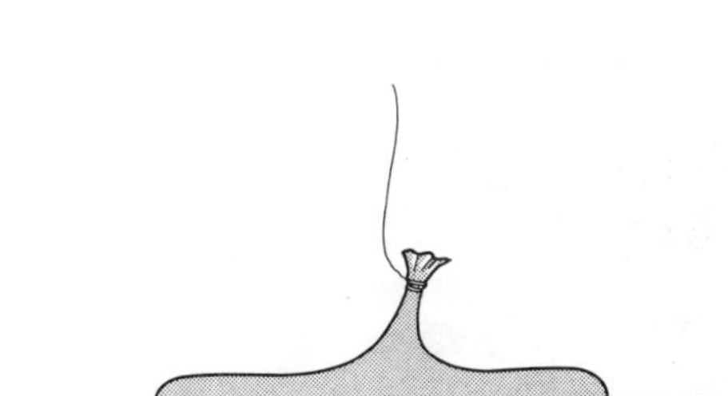

Fig. 3.11 A 'whale' on land

IN YOUR NOTEBOOK

✻ Write a heading: *Why do the largest animals live in the sea?*

✻ Draw a picture of a whale.

✻ Write this statement, filling in the gap:

The largest animals live in the oceans. They do not need large skeletons because their bodies are supported by ———.

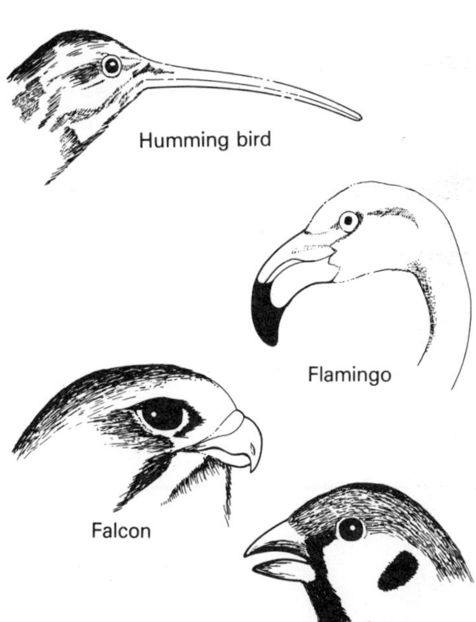

Fig. 3.12 Match these birds
with the foods listed next to them

3.3 Bony differences 2

Why do different birds have different shaped beaks? For example, look at the beak shapes in Fig. 3.12. Now look at this list of foods the birds eat. Can you match them up? (Answers on page 161.)

(a) small creatures strained from water
(b) flesh
(c) nectar from flowers
(d) seeds and insects

A bird has the kind of beak which best suits its eating habits. The shape of its beak helps it to survive.

IN YOUR NOTEBOOK

✳ Write a heading: *Variety in beak shape*

✳ Draw the birds' beaks.

✳ Under each beak write the name of the food the bird eats.

The search for food takes birds to different places. They have special feet to help in their search. For example, look at the feet in Fig. 3.13.

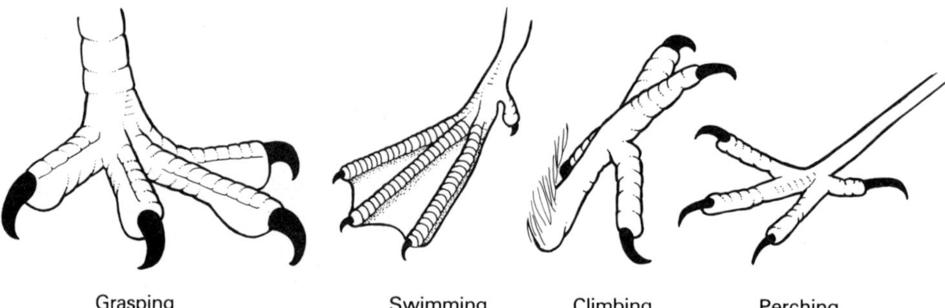

Fig. 3.13 Different birds have different feet to help in their search for food

Now look at the birds in Fig. 3.14. Can you match up the birds' heads with their feet? (Answers are on page 161.)

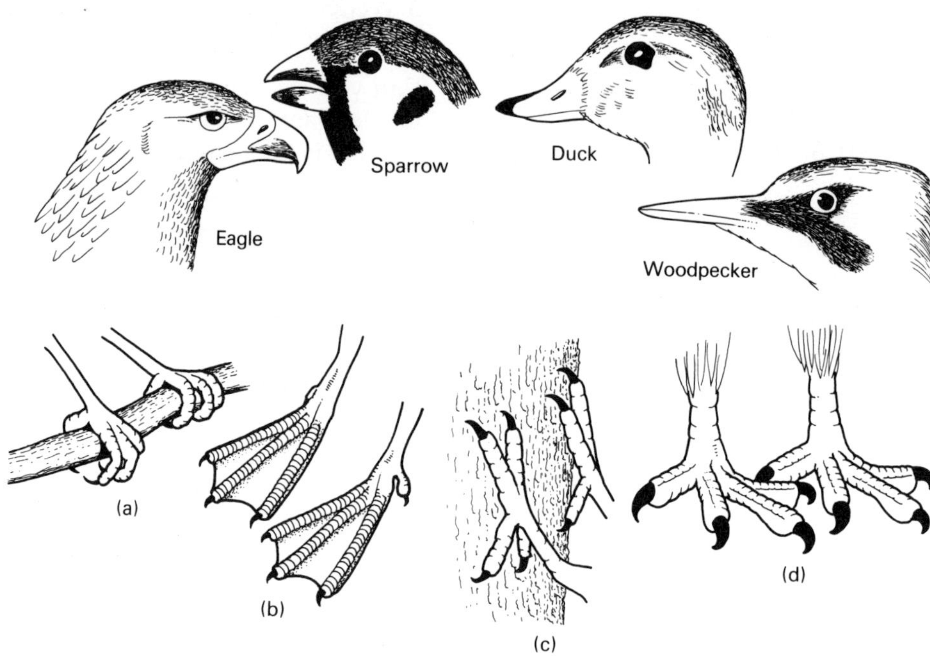

Fig. 3.14 Match the birds' heads with their feet

IN YOUR NOTEBOOK

✳ Write this statement, filling in the gaps:

The _______ and _______ of birds have special _______ to help them in movement and obtaining _______.

3.4 More about beaks and feet

The beak of a bird is very strong, and multi-purpose. It is used for feeding, building nests, fighting and preening.

Birds of prey such as the eagle, hawk, owl and kestrel have short hooked beaks (Fig. 3.15). With this kind of beak the birds can easily kill their prey and tear off flesh. They have feet with powerful claws or talons to catch and hold their prey.

Fig. 3.15 Eagle

Fig. 3.16 Woodpecker

Fig. 3.17 Great tit

The woodpecker has a sharp pointed beak for pecking away at a tree (Fig. 3.16). Two of its toes point forward and two backwards to grip the tree as it climbs. It hammers to uncover the tunnels made by wood boring insects (which it eats), and to make nesting holes.

Seed eating birds, such as sparrows, tits and finches have short beaks for picking and breaking open seeds (Fig. 3.17). Three toes pointing forward and one backwards help these birds to perch on trees.

The beaks of ducks, swans and geese act as scoops. They are flat and rounded (Fig. 3.18). Each edge has a comb-like ridge which acts as a sieve. The bird scoops up water and lets it run out through the 'sieve'. Any living things in the water are held back and eaten. The webbed feet of these birds help them swim.

3.5 Copying the birds

The people in Fig. 3.19 seem to be copying the birds. Can you say how? Can you say what kinds of birds?

On page 55 you saw that the largest animals live in the oceans. They do not need large skeletons because their bodies are supported by the water. On page 56 you saw that the feet and beaks of birds have special shapes to help

Fig. 3.18 Goose

(a)

(b)

(c)

(d)

Fig. 3.19

them in movement and obtaining food. These are just two examples of an important pattern:

Living things have many different shapes, sizes, colours, and smells. Most of them have sets of characteristics which best help them to survive.

IN YOUR NOTEBOOK

✳ Write the pattern above.

✳ Draw a box round the pattern.

✳ Try to find a picture to illustrate the pattern. Copy the picture or, if you can, cut it out and stick it into your book.

✳ How did living things come to have the best shapes, sizes, colours, and smells, for the lives they lead? Write down what you think.

3.6 The long stretch

How did the giraffe get its long neck?

The giraffe is the tallest of all animals. A fully grown male can be 5 metres tall – $1\frac{1}{2}$ metres taller than the African elephant, the second tallest animal. Of course, the giraffe gets its great height from its 2 metre long legs and its even longer neck.

Giraffes live in the grasslands of Africa (known as the savannas). They eat the leaves, twigs and fruit of scattered trees. They use their long upper lips and long tongues (over 40 cm long!) to gather food from tree branches. Giraffes, like cows, chew the cud.

Surprisingly perhaps, the giraffe has only seven neck bones – the same number as man and most other mammals.

Scientists believe that millions of years ago there were only a few kinds of mammals on the earth. Perhaps some of them looked like those in Fig. 3.21.

Fig. 3.20 Giraffe – the tallest of all animals

Fig. 3.21 Mammals that lived millions of years ago may have looked like this

Fig. 3.22 In every new generation of giraffes, those with longer necks survived and bred

All of today's mammals, including the giraffe, are descended from mammals with short necks like those in Fig. 3.21. So how did the giraffes come to have long necks?

One theory starts with the idea that the earliest giraffes had short necks. There was plenty of food at a low level on bushes and trees. But as the numbers of giraffes increased the food supply decreased. Giraffes with slightly longer necks could reach the high leaves so they had a better chance of surviving long enough to breed. They passed on their longer necks to their offspring.

In every generation of baby giraffes there were a few with longer necks. These survived and bred. The ones with shorter necks died out. Small changes built up into big changes and so it went on, for hundreds of thousands of years. Until all the giraffes had really long necks.

Giraffes, like other animals, have survived by changing as their surroundings changed. Small changes have happened almost by accident from one generation to the next.

Scientists call this process *evolution*. They have one word for 'changing to suit the surroundings'. The word is *adapting*. The process has taken hundreds of thousands of years.

Fig. 3.23 How many faces are showing?

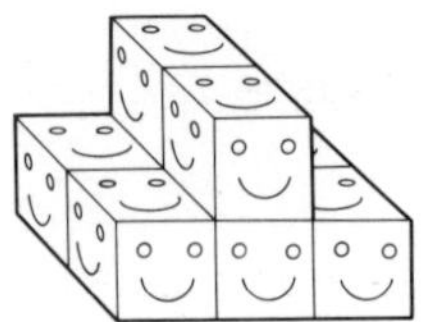

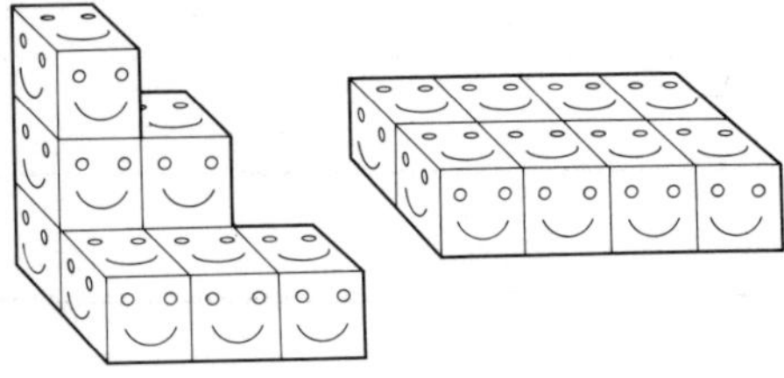

Fig. 3.24 How many faces are showing in these arrangements of blocks?

3.7 The long and the short and the fat

Scientists are interested in the ways that animals – such as the giraffe – have changed to survive. So they are interested in making patterns which connect shape and size with 'survival ability'.

The amount of skin that an animal has can make a lot of difference to its survival. For instance, animals lose heat and water through their skins. Animals have many different shapes and sizes. It is difficult to measure the amount of their skin – their 'surface area'. But you do not need real animals – you can get at the patterns using a 'model'.

Join eight blocks together as in Fig. 3.23. They take up eight times as much space as one does but how many faces are showing? Write down the number of faces.

There are other ways of joining eight blocks together to have even fewer faces showing (Fig. 3.24).

IN YOUR NOTEBOOK

✳ Write a heading: *Using a model to make a pattern about surface area*

✳ Draw the shape with the least surface area.

✳ Draw the shape with the greatest surface area.

✳ Write these statements, filling in the gap:

In the model the same number of bricks, taking the same amount of space, could be arranged in shapes to give different surface areas. Long thin shapes have a _______ surface area than short fat shapes.

Now do an experiment to study the way different shapes cool down.

Investigation 3.3 How do different shapes cool down?

You will need:
2 plastic boxes (lunch boxes will do but perspex models would be even better). They must take up the same amount of space, but one must be long and flat and the other chunky. They must have a hole to pour hot water into and another with a thermometer in.
corks to lift the boxes off the ground
a measuring cylinder

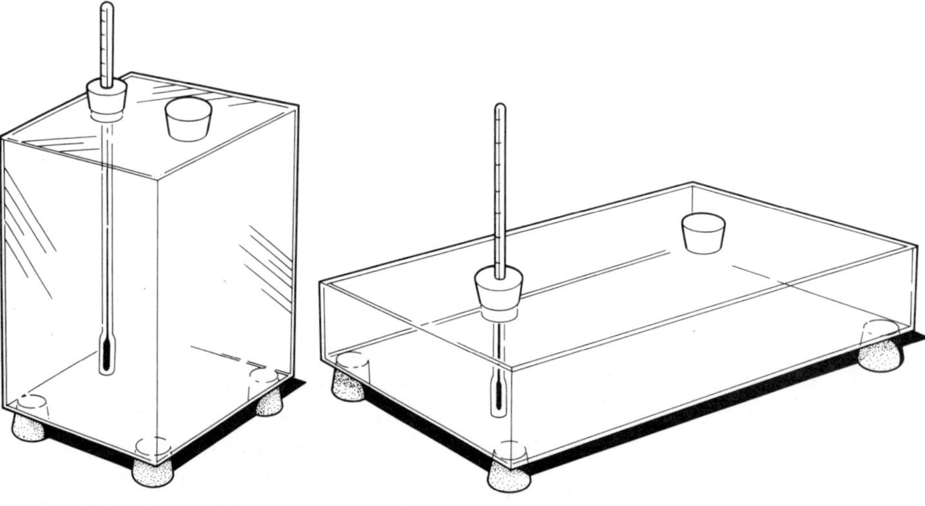

Fig. 3.25 Investigation 3.3

1 Use the measuring cylinder to fill both boxes with the same amount of hot water and replace the bungs.

2 Take the temperatures at the beginning and then every five minutes.

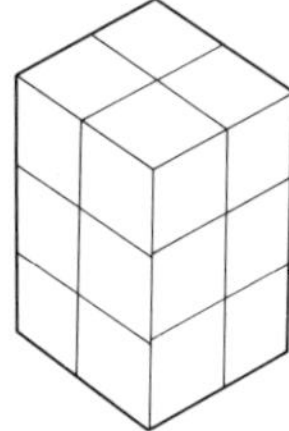
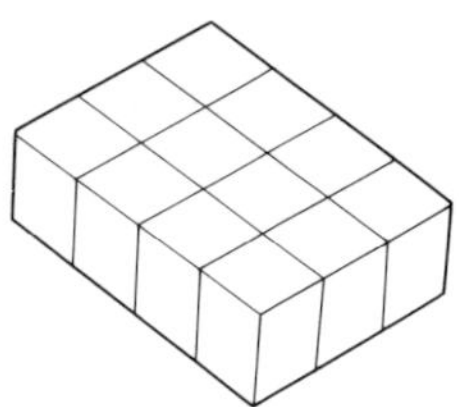

Fig. 3.26 If you heated these two shapes, which one would cool down the fastest?

IN YOUR NOTEBOOK

✳ Write a heading: *An experiment to study the cooling of different shapes*

✳ Draw a diagram which shows how you did the experiment.

✳ Draw up a table like the one below. Fill in your results. Call the boxes A and B.

Time/mins	0	5	10	15	20	25	30
Temperature of A/°C							
Temperature of B/°C							

✳ Write this statement, filling in the gaps:

The ______ the surface area the ______ the shape cools down.

3.8 Adapting?

So we know that chunky shapes have less skin showing than skinny shapes. Elephants have a small surface area for their chunky shape. They live in very hot countries. They have big ears to help keep themselves cool. Their ears help to increase their surface area.

What sort of country do you think the fox in Fig. 3.27 lives in? Why? What reason could you give to support your answer? What about the fox in Fig. 3.28?

Fig. 3.27

Fig. 3.28

3.9 . . . or not adapting?

You wouldn't expect to see a penguin on a hot tropical island or a camel in Greenland – they are adapted to live in other climates. And they are very well adapted indeed. Compared with them, human beings seem hardly to have adapted at all.

In fact, of all the animals in the world, humans are the least well adapted to their surroundings. So how do we survive? The answer is that of all the animals in the world, humans are the cleverest. We have learned to adapt our surroundings to suit ourselves.

Fig. 3.29 Double glazing and air conditioning (the boxes under the windows) are two ways we humans can adapt our surroundings to suit ourselves

We can control our heat loss by wearing clothes. We can make the places in which we live hot or cold, wet or dry as it suits us. We can control our environment.

Some scientists think that we are getting so good at changing the world to suit ourselves that we will not need to adapt or evolve any more! Do you think that the power of humans has made the world a better or worse place? How do you think that humans might evolve in the future?

3.10 Penguin patterns

Penguins are found in many different parts of the world. They are all roughly the same shape but come in very different sizes! For instance, take the five characters in Fig. 3.30.

The five penguins show clear signs of adaptation. (Remember – that means they have changed slowly over thousands of years to suit their surrounding. Also remember – page 61 – big fat shapes lose heat more slowly than small skinny shapes.)

Remembering these things, can you say which penguin lives in which of the places marked on the map in Fig. 3.31?

IN YOUR NOTEBOOK

✶ Write a heading: *Penguin patterns*

✶ Draw a map (trace it first).

✶ Make small drawings of the penguins in their right places.

✶ Don't forget to show the temperatures in the different places.

Fig. 3.30 Five different species of penguin

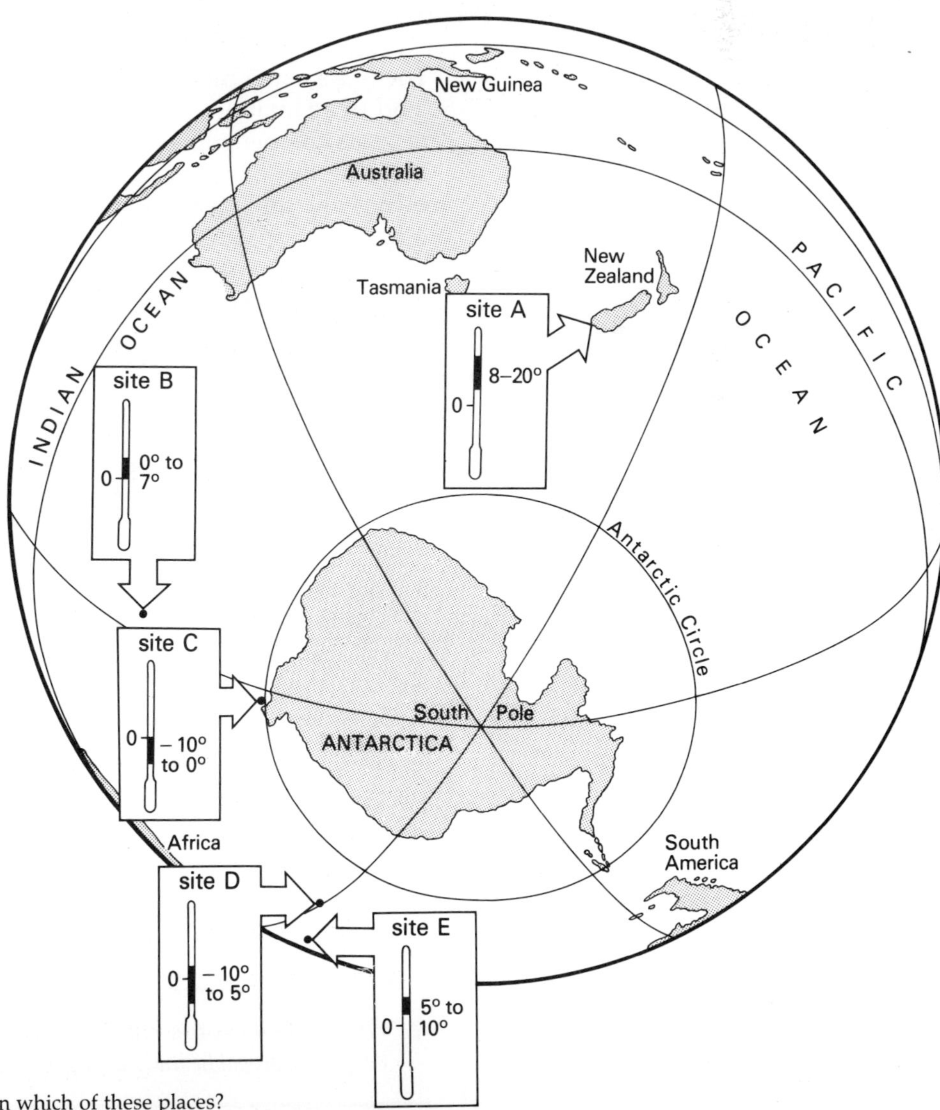

Fig. 3.31 Which of the penguins in Fig. 3.30 lives in which of these places?

Fig. 3.32 How does huddling together help these penguins?

There are some more penguins in Fig. 3.32. They are showing a different kind of adaptation. Scientists call it behavioural adaptation. Over thousands of years the penguins have picked up special ways of behaving. For instance, those in the picture prefer huddling together to standing on their own. How do you think that helps them? You can do an experiment to find out.

Investigation 3.4 Why do penguins huddle together?

You will need:

15 test-tubes
2 thermometers (–10 to 110 °C)
1 elastic band

1 stop clock
something to hold the tubes
a supply of hot water

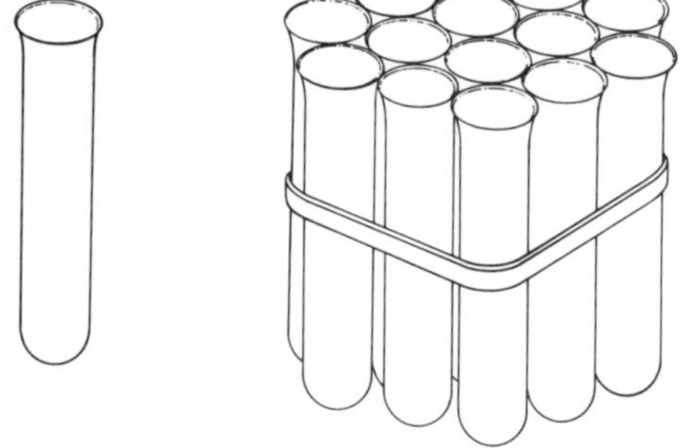

Fig. 3.33 Investigation 3.4

1 Put one test-tube by itself and the others into a huddle held by the elastic band.
2 Fill all the test-tubes with hot water.
3 Take the temperature of the water in the single test-tube. Write it down.
4 Take the temperature of a test-tube right in the middle of the huddle. Write it down.
5 Leave the test-tubes for ten minutes, then take the temperatures again. Do the same after another ten minutes.

IN YOUR NOTEBOOK

✶ Write a heading: *An experiment to study cooling in a huddle*

✶ Draw a diagram of the experiment.

✶ Write this statement, filling in the gap:
The test-tube ______ lost heat most quickly.

✶ Write a few words starting with:
Penguins probably huddle together because ______

3.11 Choosing the best

All of our domestic animals (dogs, cats, horses, cows, etc.) are descended from wild breeds. The breeds in Fig. 3.34 were popular in the Middle Ages.

Fig. 3.34

(a) Highland cattle

(b) Chillingham wild bull

(c) British Whites

Three breeds of cattle found on today's farms can be seen in Fig. 3.35.

(a)

(b)

(c)

Fig. 3.35 Three breeds of cattle on British farms today
    (a) Hereford
    (b) Hurwood
    (c) Friesian

IN YOUR NOTEBOOK

✳ Write a heading: *Breeding the best animal for the job*

✳ Sketch the three breeds.

✳ Write out these statements, filling in the gaps (a bit of research is needed for this):

The breed which gives the largest amount of milk is: ______.
The breed which gives the creamiest milk is: ______.
The breed which gives the best beef is: ______.

Sometimes people breed new varieties of animal or plant just for the sake of it, rather than for any useful reason. Why should they want to do this?

Fig. 3.36 A variety of rose called 'Anna Ford'

Man can choose the things he likes about different types of animal. He can breed the animals together so that features he likes are passed on from the parents to the calves (so the calves inherit the right variations). Fig. 3.37 shows how this is done.

Fig. 3.37 A breeder selects the animals or plants that show the variation best . . .

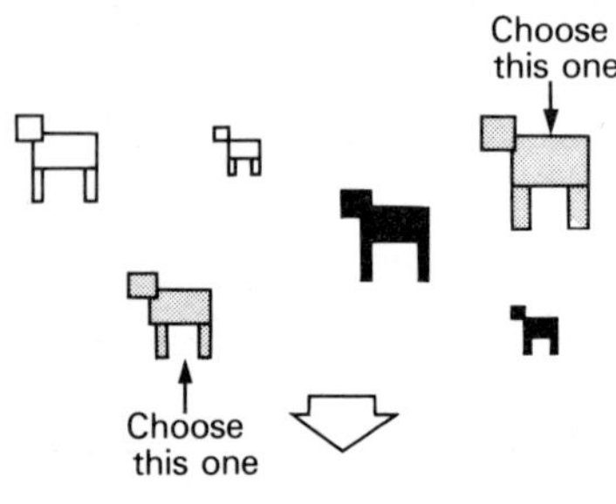

. . . and breeds them together.

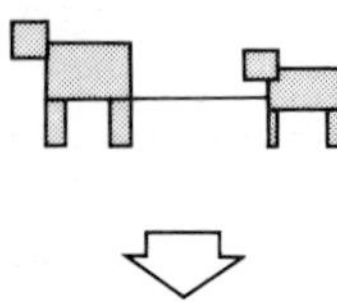

He chooses the best offspring from these two . . .

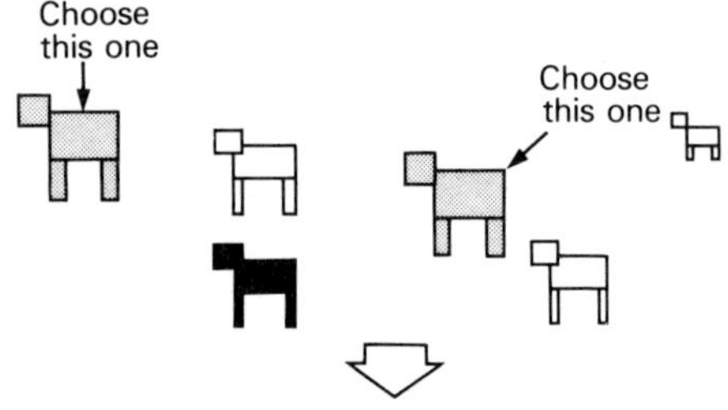

. . . and breeds them together.

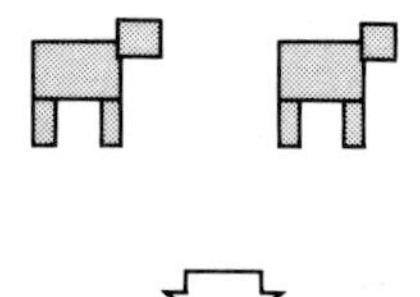

And so on . . . until the animal or plant is exactly what he wants.

IN YOUR NOTEBOOK

✳ Try to find a picture which shows a good example of selective breeding (a plant or an animal). Paste it into your book.

✳ Under the picture, write this statement:

Man makes use of inherited variation. He breeds together a male and a female both with the characteristic he is looking for.

There is always a danger that the selectively bred stocks might be hit by disease or bad weather. If things were really bad, the stocks could be wiped out. So it is important that the wild breeds are preserved – just in case. As you will see, the 'plans' for new plants and animals are stored in seeds and sperm. 'Banks' of these are kept in deep freezes.

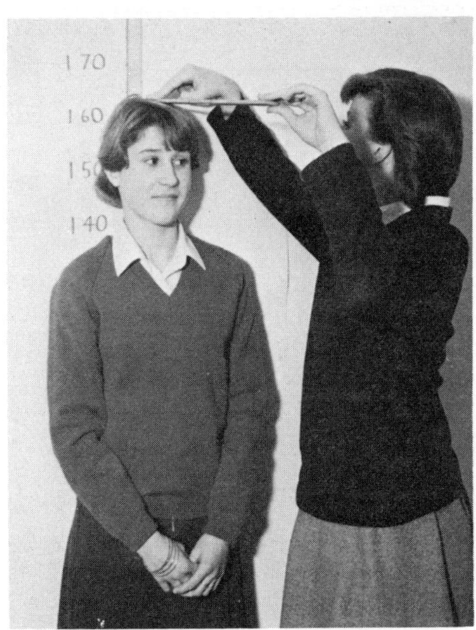

Fig. 3.38 Measuring a person's height

3.12 How do humans vary?

Now look at one way that the people in your class differ from each other.

Investigation 3.5 Measuring people's heights

You will need to measure people without their shoes on. Ask a friend to help you. Fig. 3.38 shows one way of doing it.

IN YOUR NOTEBOOK

✻ Draw up a table like this. Fill in ticks for each of the people in your class.

Person's height in centimetres

Name	130 to 134	135 to 139	140 to 144	145 to 149	150 to 154	155 to 159	160 to 164	165 to 169	170 to 174	175 to 179	180 to 184	185 to 189
Gerry										∨		
Bobbie							∨					
Richard					∨							
Rose				∨								
Jerry							∨					
Pete					∨							
Mick					∨							
Frances						∨						
Totals for this sheet	0	0	0	1	3	1	2	0	0	1	0	0

Ask your teacher if there are any 'data sheets' like this from other classes that you can use.

✻ Draw a special chart (called a histogram) to show the heights of all the people you have data for. The scales of your chart should look like Fig. 3.39. Colour in one square box for every person whose height comes in each range. Can you see any patterns in the histogram you have drawn?

✻ Write a few words starting with:

In a group of people of the same age there is ______.

If you had time you could look for patterns by drawing histograms for other measurements (arm span for instance).

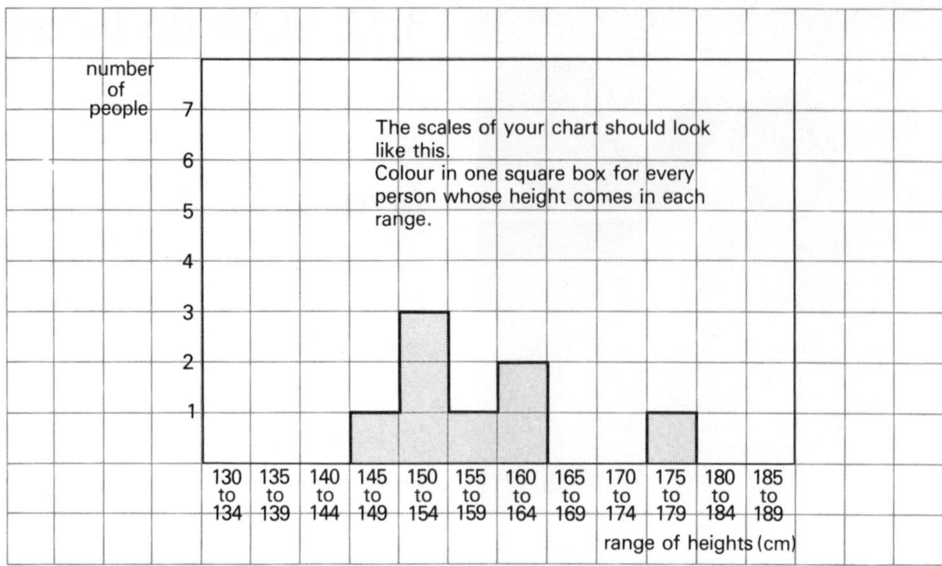

Fig. 3.39 Graph of numbers of people in different height ranges

3.13 More variation

Fig. 3.40 shows people who can roll their tongues and one person who is trying to but is unable to do it. Are you a tongue-roller or a non tongue-roller?

Fig. 3.40

Investigation 3.6 How many people in your class can roll their tongues?

IN YOUR NOTEBOK

IN YOUR NOTEBOOK

* Write these statements, filling in the gaps.
In my class there are _______ *tongue-rollers.*
There are _______ *non tongue-rollers.*
What would a histogram for tongue-rolling look like?

* Write a few words about it.

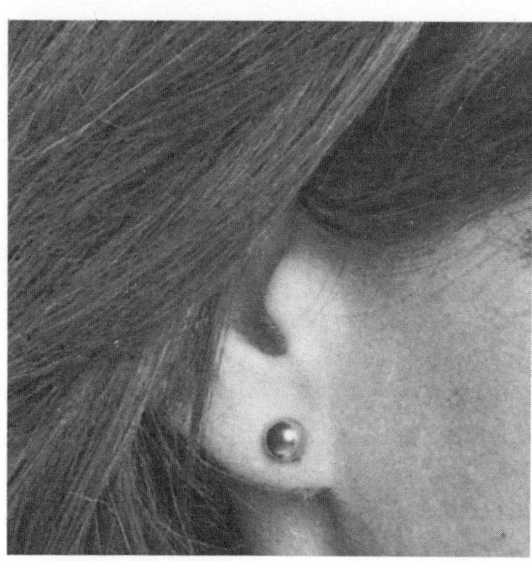

(a)

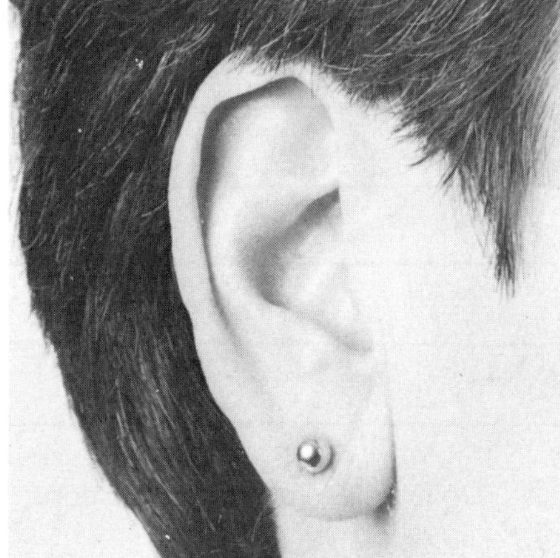

(b)

Fig. 3.41 (a) This ear lobe is attached
         (b) This ear lobe is unattached

The ear lobe in Fig. 3.41(a) is attached. The ear lobe in Fig. 3.41(b) is unattached.

Investigation 3.7 How many people in your class have attached ear lobes?

IN YOUR NOTEBOOK

✻ Write these statements, filling in the gaps:

In my class there are _______ people with attached ear lobes. There are _______ people with unattached ear lobes.

In the next chapter you will read about the reasons for variation in humans – the reasons why everybody is different.

3.14 Hey . . . What do you think?!

People vary in the way they *think* about things too.

The only way to discover and compare people's ideas is to ask them!

To do that in a scientific way, you need to get organised. You need to think out all the questions you want to ask. Then you need to put them in a list. Such a list of questions is called a *questionnaire*.

Fig. 3.42 A street survey

Investigation 3.8

Here is a questionnaire. There are some spaces for your own questions.

1 How often do you play football at weekends? How many weekends in a year?

2 How often do you go to a youth club (in a week)?

3 How often in a year do you read a book of your own choice?

4 How often in a week do you read a newspaper?

5 How many hours in a week do you watch TV?

6 Which radio station do you prefer listening to?

7 What time do you wake up on a weekday morning?

8 What time do you usually go to bed on a weekday?

One of the 'rules' about questionnaires is: 'Do not talk about the questions with anyone else before you answer them'.

Can you see the reason for the rule? Keep to it as you fill in your answers. When you have finished, get together with your friends and compare answers. You could give the questionnaire to other groups of people – in other classes for example.

IN YOUR NOTEBOOK

✶ Write a heading: *How do people's habits vary? (Using a questionnaire)*

✶ Copy out your questionnaire.

✶ Collect as many results as you can.

✶ Show the results on histograms.

Why not have a go at making up your own questionnaire on any subject you like – music, food, sport, politics, for example. Try it out on your friends, parents and relatives.

IN YOUR NOTEBOOK

✶ Try to answer these questions:
  What kinds of people might want to use questionnaires? What might they be trying to find out? Why might they be doing it?

Checkout

Keywords

adaptation
behavioural adaptation
environment
evolution
histogram
inheritance
organism

questionnaire
selective breeding
surface area
variation
variety
volume

Patterns

1 Organisms vary in their shapes, sizes, colours, and smells.
2 Organisms have sets of physical characteristics which best help them survive.
3 The surface area of an object depends both on its shape and on its size. Tall thin objects have larger surface areas than short fat objects.
4 Species have changed to suit their changing surroundings (they have adapted). These small changes have taken place over millions of years.
5 Organisms have evolved ways of behaving which help their success.
6 Man uses inherited variation to breed animals and plants which have the best characteristics for his needs.

True or false?

1 The largest animals alive today live in the oceans.
2 Animals which live in the oceans have bulkier skeletons than animals of a similar size on land.
3 Whales need smaller skeletons than elephants because their bodies are lighter.
4 The shape of a bird's beak can tell you how it lives.
5 The smaller the surface area of an object the faster it cools down.
6 A cube with one metre edges has a volume of 6 m^3.
7 Most warm-blooded animals living in hot climates have small surface areas for their volumes.
8 Penguins are better adapted to very cold climates than Eskimos.
9 If everyone's feet were the same size you would get a histogram like the one in Fig. 3.43.
10 Prize bulls can produce new offspring long after they are dead.

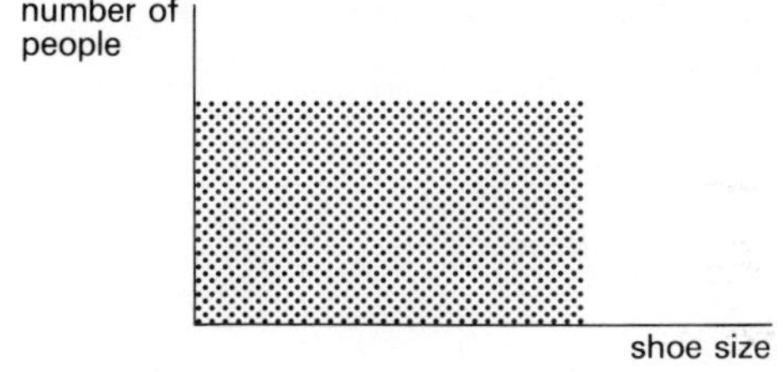

Fig. 3.43

Problems

1 *Read this passage about the three-toed tree sloth and try to answer the questions.*
The odd life of the sloth
Hanging upside down high in a Cecropia tree he sleeps most of the day and night. He lives on Cecropia leaves – no other animal will touch them. Green algae grow all over his fur and green caterpillars graze on it. He never wanders about in the trees. Only once a week he comes slowly down to the forest floor to defecate and urinate. He always goes to the same spot, finding it by its smell.

(a) How has the body of the sloth adapted to suit its surroundings? (Write down two ways it has adapted.)
(b) How do the behavioural adaptations of the sloth protect it from attack?
(c) No one is quite sure how male and female sloths get together to mate. How do you think they find each other?

2 *Read this passage about dinosaurs and try to answer the questions.*
The giant dinosaurs lived successfully on the Earth for about 100 million years. (Man has been around for less than 5 million!) Around 60 million years ago, the dinosaurs suddenly disappeared – no one is quite sure why. There are several theories about events at that time, for instance . . .

Theory 1: says that changes in the shape of the Earth's crust caused many shallow inland seas to drain away. (The giant dinosaurs lived in the swamps at the edges of these seas.)

Theory 2: says that the climate of the Earth changed and it became much colder.

Fig. 3.44 The World when the dinosaurs ruled – as seen by a French artist in 1881. In those days people believed all dinosaurs were like crocodiles.

(a) Why would it be difficult for the giant dinosaurs to live on dry land? (Try to give two reasons.)
(b) Do you think it would matter to a dinosaur if its surroundings became colder? Explain your answer.
(c) What do you think happened to the dinosaurs? Write a short story called *'The Last Days of the Dinosaurs'*.

4 Cell

4.1 Biomaterials

In the last chapter you looked at variety in collections of things – collections of human beings, collections of penguins, perhaps even collections of dandelions! Now you are going to look at a different kind of variation – variation in the stuff that living things are made of. You can call the different kinds of stuff *biomaterials*.

Biomaterials can be hard. The enamel on your teeth is one example. Can you think of others?

Biomaterials can be waterproof. Leaves are often covered with a transparent waterproof skin made from stuff called cutin.

Biomaterials can be soft. Fur and hair are made of keratin – a material formed from dead skin cells. Hooves, horns, and beaks are made of keratin too.

Fig. 4.1 A locust laying its eggs. The whole of the back part of its body including the egg-laying tube has pushed into the sand.

Biomaterials can be elastic. The egg-laying tube of a locust can stretch to fifteen times its normal length and spring back completely. The best that man-made elastic can do is to stretch to six times. And, unlike man-made elastic, the locust's material doesn't tighten up as it stretches. Knickers or underpants, body-stockings or spacesuits made from locust elastic would feel no tighter however large the wearer.

So – guess what – scientists are trying to find out how the locust does it and what its elastic is made from, so they can copy it.

Fig. 4.2 This biomaterial is found in your body. What is it?

Fig. 4.2 shows one of the biomaterials in your body. Can you guess what it is?

Did you guess? The picture shows a very thin slice taken across a leg bone. It was photographed through a microscope. The big black spots are where the slice has cut through tubes in the bone. The tubes carry nerves and blood vessels through the bone. Fig. 4.3 will help to make the photo clear.

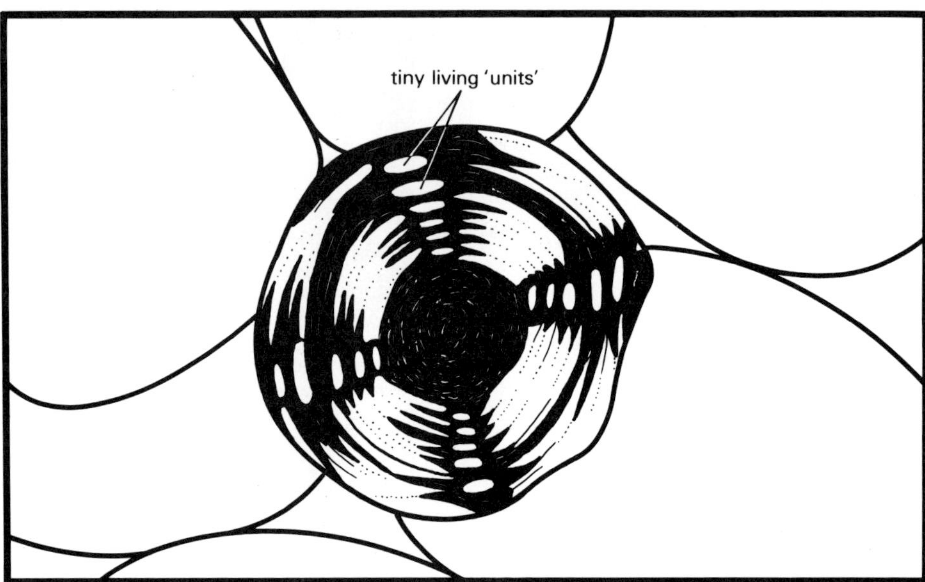

Fig. 4.3 Drawing of Fig. 4.2

4.2 Tissues

Bone is as strong as reinforced concrete, as hard as granite, and much, much lighter than both. It is flexible and springy. Its secret lies in its make-up. Most of it is made of threads of a protein called *collagen* cemented together with minerals such as calcium phosphate. You can find out what collagen and calcium phosphate are like.

Investigation 4.1 *What are collagen and calcium phosphate like?*

You will need:
chicken bones
dilute hydrochloric acid
Bunsen burner, tripod and gauze

1 Put a chicken bone into dilute hydrochloric acid for a few days. The calcium phosphate will dissolve leaving the collagen.
2 Burn a bone on a tripod and gauze over a Bunsen flame. (The smell is awful!) You will be left with the calcium phosphate.

Look again at the microphotograph of the bone slice. Can you see the dark regions spreading out like the spokes of a wheel from the supply tubes? These too are spaces. They contain the tiny units which build and maintain the bone, and the thousands of tiny tubes they send out to help. (See Fig. 4.3.) The units don't show up in the photograph (Fig. 4.2) because they are transparent to the light used to take the picture.

Most biomaterials are made up of groups of little units. Can you remember what the little units – the building blocks of living things – are called?

The little units are called cells. Biomaterials made up of teams of similar cells working together are called *tissues*.

IN YOUR NOTEBOOK

✳ Write a heading: *Living building blocks*

✳ Write out these statements, filling in the gaps:

The body of an animal or plant is made up of millions of tiny building blocks called ________.
Some ________ *are the same, some are different. Some of the same kind join together in groups and work as a team. The materials they form are called t* ____

4.3   A close look at cells

Investigation 4.2 *Looking at some cells under a microscope*

You will need:
a leaf   a microscope   an onion
nail varnish   4 microscope slides

1 Scrape some cells from the inside of your cheek with a clean finger.
2 Paint a thin layer of nail varnish on the outside of the leaf. Wait until it dries then peel it off. It will pull off some leaf cells with it.
3 Tear a thin layer of skin from the outside of the onion.
4 Smear some blood on to a microscope slide. (Blood from fresh meat or from your teacher!)

Put each sample on to a clean slide and (except for the blood) put a tiny drop of water on top of it. Look at your samples through the microscope.

What did you see? Anything like the cells in Fig. 4.4? The cells have been drawn roughly the same size. Were they all the same size? If not, which was the largest? Which was the smallest?

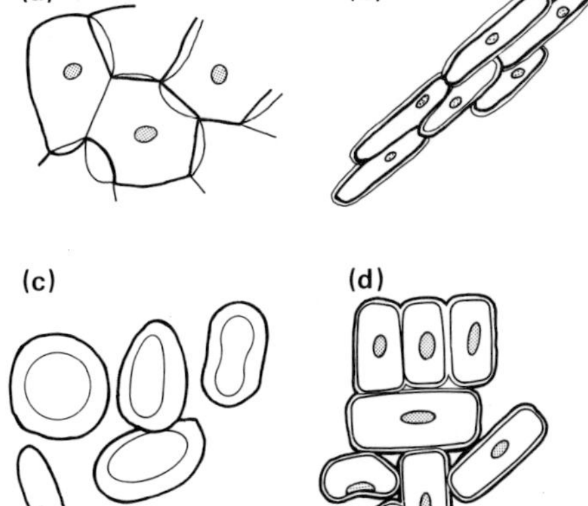

Fig. 4.4 (a) Cheek cells
(b) Onion cells (from the outside layer called the epidermis)
(c) Red blood cells
(d) Leaf cells (from the epidermis)

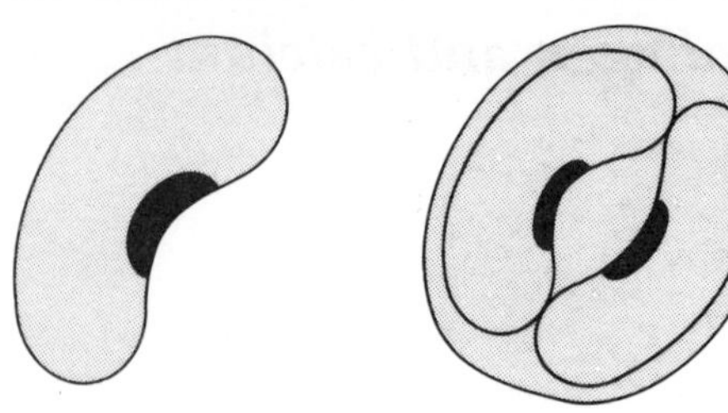

Fig. 4.5 Guard cells

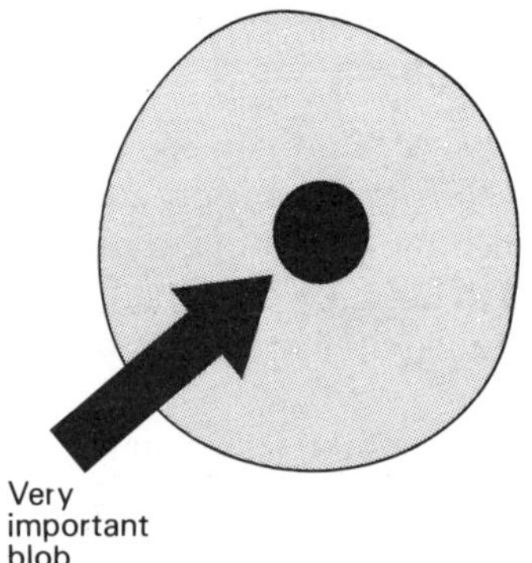

Fig. 4.6 The cell nucleus

IN YOUR NOTEBOOK

✳ Draw diagrams of the four kinds of cell.

✳ Put these labels on your drawings:
  cheek cells   red blood cells   leaf cells (epidermis)
  onion cells (outside layer called the epidermis)

If you saw something like Fig. 4.5 when you looked at leaf cells you were looking at a special cell which lets water and air out of the leaf. It is called a *guard cell*. Two of them fit together to build part of a leaf.

The small blob that you can sometimes see in a cell is very important. It is called a *nucleus*. The nucleus tells the cell how to act and it carries the inheritance 'plans' (see page 66). Almost all cells have a nucleus but it is often hard to see.

Fig. 4.7 shows a drawing of a cell magnified many times. As you will see, the cell is like a little community – full of different bits doing different jobs.

If you look at pictures like Fig. 4.7 for a while, you start to imagine that cells are flat – like pancakes. You need to remind yourself that they are more like little blobs.

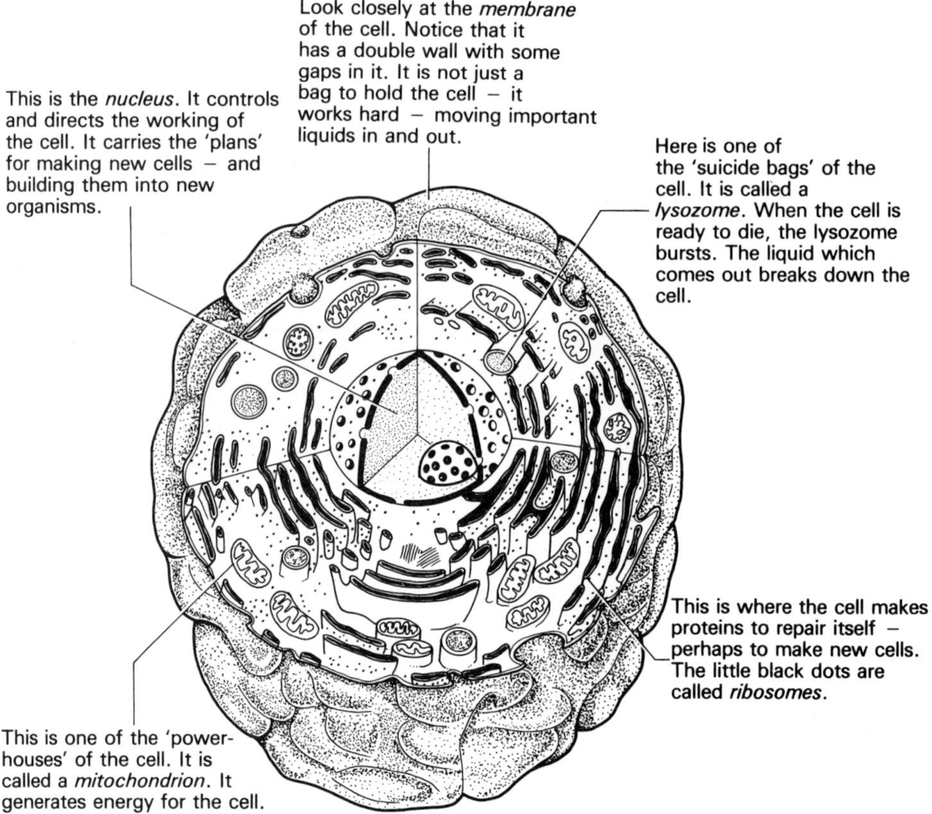

Fig. 4.7 Drawing of a cell magnified about one thousand times

A plant cell has extra parts:
— A wall made of stuff called *cellulose*. The cellulose supports the cell and gives it a better shape – but it stops the cell moving. Animal cells, which do not have a cellulose wall, can move about.
— A space often filled with watery stuff called *sap*. The space is called a *vacuole*.

4.4 Why are you like you are?

Look at yourself in a mirror. Perhaps you look a bit like your parents? Or like your grandparents? But you do not look exactly the same as any of your relatives.

Why do you look the way you do? The answer is to do with some very special cells. You began your life when a sperm cell (called a male *gamete*) joined up with an egg cell (called a female *gamete*). The result was a fertilised egg (called a *zygote*). To start with, the zygote was only the size of a full stop. Soon it began to grow into an embryo inside your mother's womb (*uterus*). After about 280 days you were ready to be born and to begin your life in the world outside. Quite a lot happened in the 280 days. Fig. 4.9 shows three of the stages.

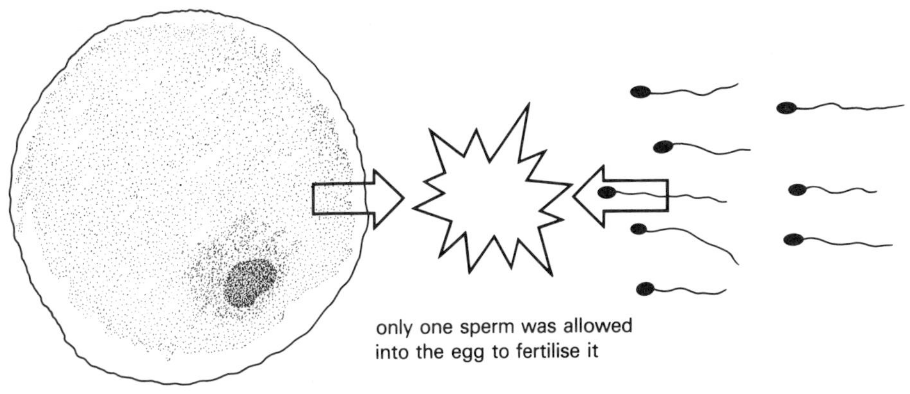

Fig. 4.8 Fertilisation

Fig. 4.9 Three stages of development in the uterus

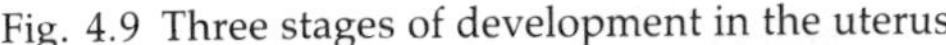

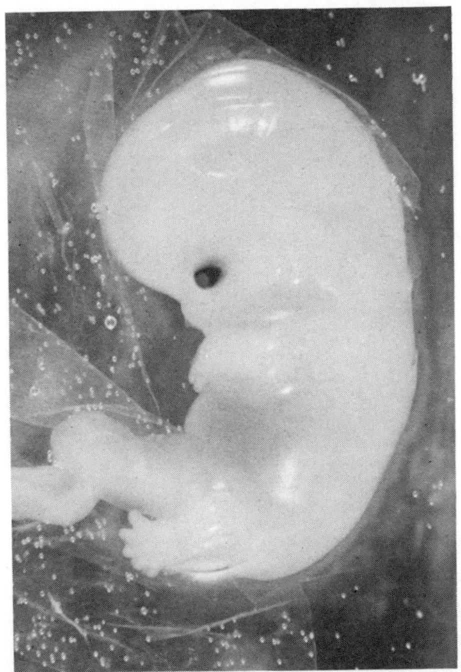

(a) At 5 weeks

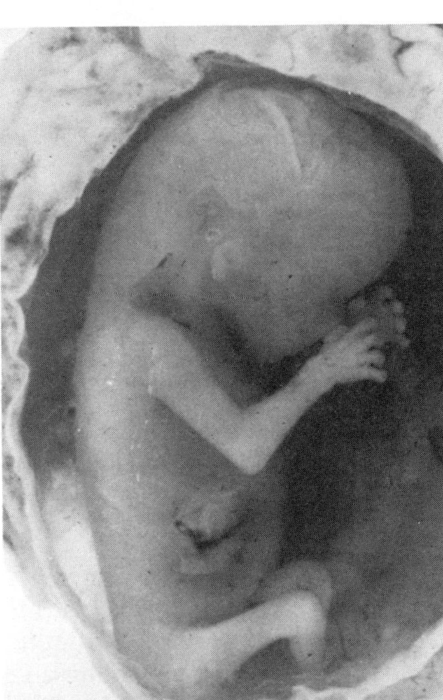

(b) At 12 weeks

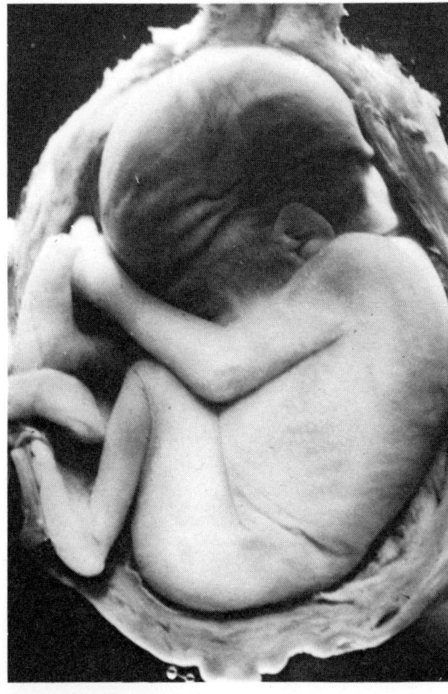

(c) At 20 weeks

How do twins start off? How does their start differ from a single baby's? You probably know that there are two sorts of twins: *identical* twins and *fraternal* twins. They begin in different ways.

Fig 4.10

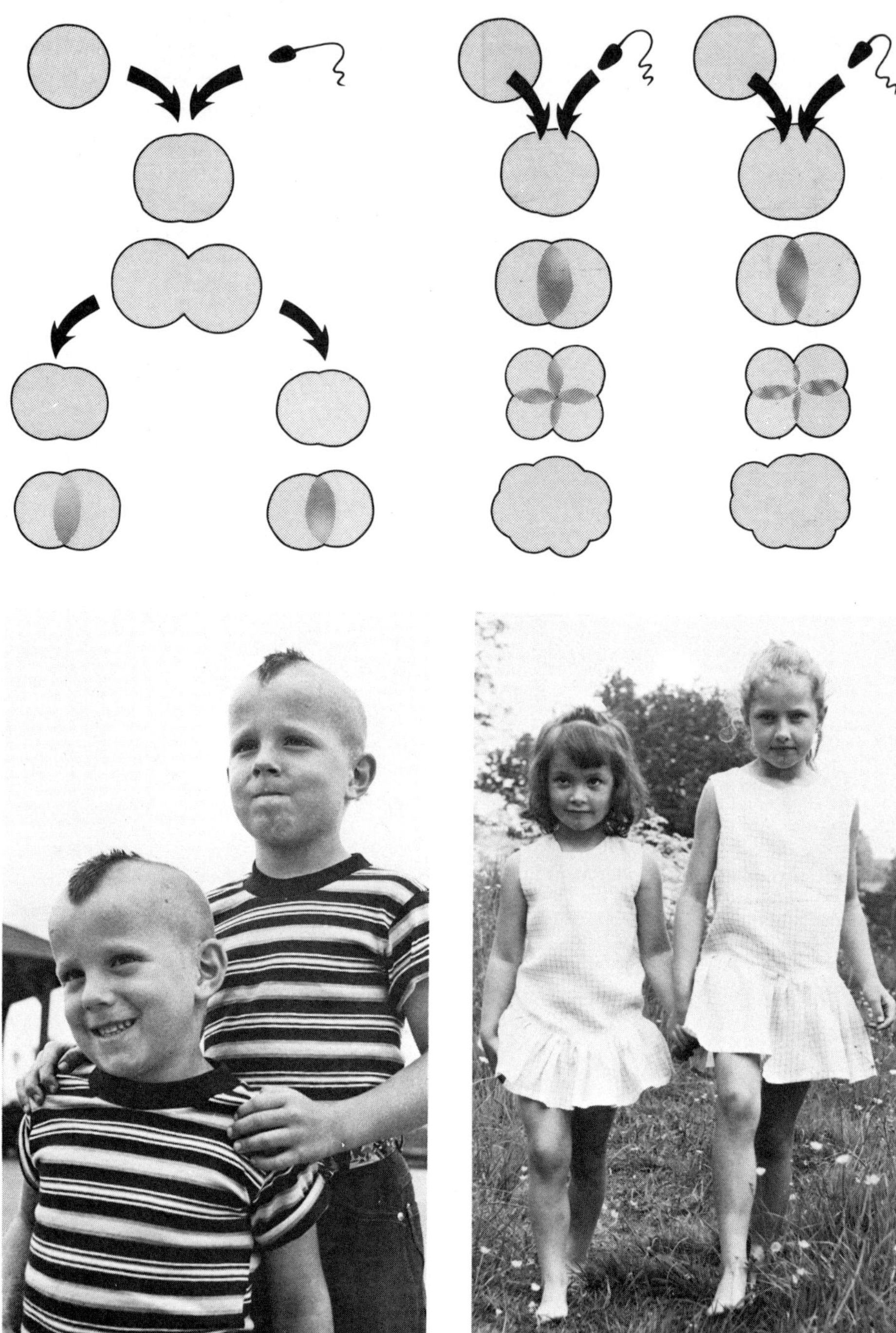

(a) Identical twins

(b) Fraternal twins

Identical twins start when one sperm fertilises one egg. The zygote splits into two. Each half develops (as the cells divide) into a *foetus*.

Fraternal twins develop from *two* eggs and *two* sperms. Two zygotes are produced. Each develops into a separate foetus.

Scientists have one word for 'plans for building a new human being'. The word is *inheritance*.

IN YOUR NOTEBOOK

✳ Write a heading: *A human being is a mixture*

✳ Copy the diagram on page 78.

✳ Write out these statements, filling in the gaps:

The 'plans' for making an individual person are carried partly by the ______ cell and partly by the ______ cell.
So everybody gets part of their ______ 's 'plans' and part of their ______ 's.
Identical twins develop from the same sperm and egg cells and so have the ______ 'plans'.

4.5 What other things shape a person?

Sometimes identical twins are separated at birth. Once or twice they have been mixed up with other babies in the same hospital! So they begin with the same 'plans' (inheritance) but grow up in different surroundings.

Scientists can study these twins as they grow up. They can measure things like height, mass, arm span and length of head. There seems to be a greater difference in these measurements for twins growing up in two different places than if they had grown up in the same place.

If a dandelion plant is cut up into equal bits, each bit will grow into a new plant and each will have started with the same inheritance. Any differences you can see in them at the end must be because of the environment they were grown in.

Checkout

Keywords

cell	nucleus
gamete	organ
guard cell	tissue
inheritance	vacuole
membrane	zygote

Patterns

1 Organisms are made up of tiny 'building blocks' called cells.
2 Cells contain a number of smaller units (organelles) each with a particular job.
3 Plant cells are different from animal cells. They are usually larger. They have a cellulose cell wall outside the membrane. They have a vacuole containing a solution of salts and sugars.
4 Tissues are groups of similar cells working together.
5 The cells of one tissue differ in shape from those of another.
6 The 'plans' for making an individual person are carried partly by the egg cell and partly by the sperm cell.
7 The way a person develops depends both on the 'plans' inherited from the parents and on the environment.

True or false?

1 It takes at least two cells to make an organism.
2 All cells have nuclei.
3 All the energy needed to keep a cell going comes from the nucleus.
4 The cell membrane can control the movement of substances into or out of the cell.
5 A plant cell has a cell wall instead of a membrane.
6 'Zygote' is just another name for a human egg cell.
7 All the information needed to build a new human being is contained in an egg cell.
8 An egg cell does not divide until it has been fertilised.
9 Fraternal twins develop from two fertilised eggs.
10 A complete new plant can grow from a piece of plant tissue.

Problems

1 Fig. 4.11 shows a picture of a set of quads. Are these quads 'identical' or 'fraternal'? Look at the diagrams on page 78. Draw a similar diagram to show how quads develop.

Fig. 4.11 Quads

2 *Read this passage and try to answer the questions:*
About a hundred years ago, in the famous wine district of Medoc in France, the crops of grapes were being hit by a deadly disease. Thousands of pounds worth of damage was being caused by a microscopic single-celled plant – a fungus – called downy mildew.

To reproduce, this fungus produces a special growth on its outside. When wet, the growth lets go hundreds of tiny things called zoospores. These settle down on the leaves of a plant and send out many fine threads. Some of the threads find their way into the leaves. There, they break down the cells to make food substances for the fungus. They spread so quickly that the leaves soon die.

One day in 1882, a scientist called Pierre Millardet was thinking about the problem. Out on a walk, he noticed that the vines near the road were much healthier than those well into the field. He knew that the farmers were in the habit of sprinkling 'Bordeaux Broth' – a mixture of copper sulphate and lime – on to the grapes near the road, to give them a nasty taste.

Walking home, Pierre remembered that, some years before, he had found that the fungus would not produce spores in the water from his kitchen pump – though it would in dew or rainwater. He remembered that the supply tube to his pump was made of copper. Suddenly he thought he had the answer!

(a) How do you think the threads from the spores find their way into the vine leaves?
(b) Why did the farmers put Bordeaux Broth on the grapes near the road?
(c) What was Millardet's answer to the disease?
(d) Suppose you had been Millardet's assistant. What would you have advised him to do next? What experiments would you have suggested?

5 Molecule

5.1 Air – is it there?!

How do we know that air is around us? There are some clues in Fig. 5.1.

Fig. 5.1

IN YOUR NOTEBOOK

* Write a heading: *Evidence for the air*

* Write a letter to an imaginary six-year-old to explain what we know about air and how we know it.

* You have probably seen an experiment like that in Fig. 5.2. You need a roomful of people – most with their eyes closed – and a bottle of ammonia. After the stopper is pulled out of the bottle, people put their hands up and open their eyes when they can smell ammonia. It is not very long before the ammonia gas has reached the back of the room. If you have a stop watch you can even time it and work out the speed of the gas. Can you say how you would do that?

5.2 Diffusion

The ammonia experiment is very simple but, if you think about it, the results are amazing!

The air in the room and the ammonia in the bottle seem quite still, yet the ammonia moves through the air to the back of the room.

Fig. 5.2 How long does it take for the smell of ammonia to reach the back of the room?

This property of gases to spread out, moving through and mixing with other gases, is called *diffusion*. Diffusion is a very important idea in science. You can investigate it using two boiling tubes and a supply of hydrogen gas. You will need to remember that *when mixed with the air, hydrogen gas lights with a pop.*

Investigation 5.1 Diffusion

You will need:
2 test-tubes
a splint
a source of hydrogen gas

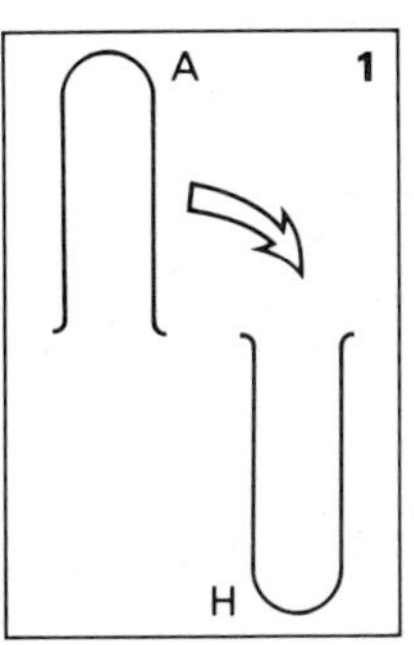

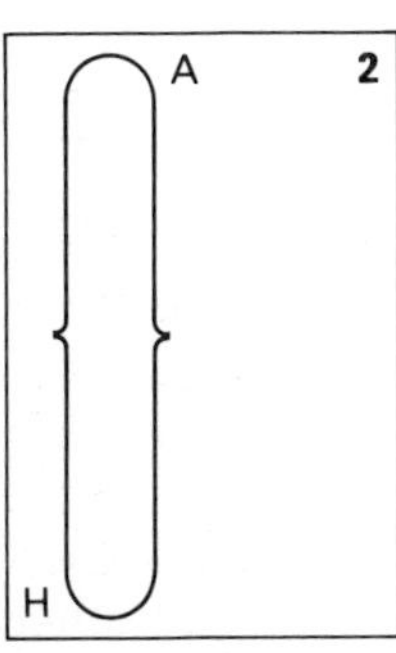

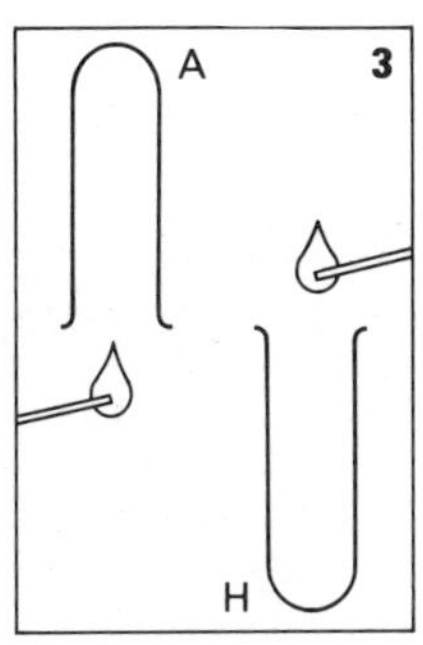

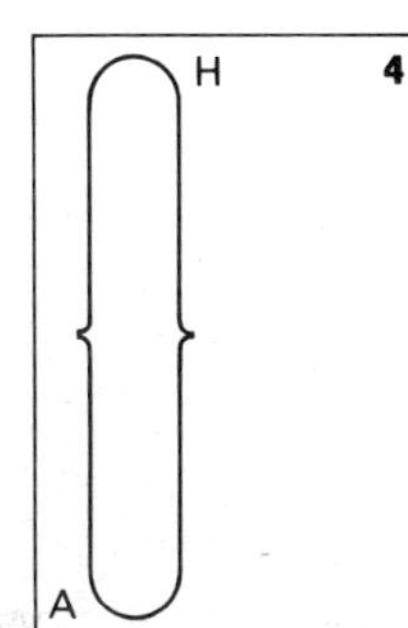

Fig. 5.3 Investigation 5.1

1 Fill a test-tube with hydrogen gas (tube H). The gas is lighter than air so you must fill the tube from the bottom upwards.
2 Link the tube to a tube full of air (tube A).
3 Wait a minute or so, then test both tubes with a lighted splint.
4 Repeat steps 1 and 2, this time with the hydrogen tube (tube H) on the top.
5 Again, test both tubes with a lighted splint.

IN YOUR NOTEBOOK

✷ Write a heading: *Hydrogen diffusion*

✷ Draw some diagrams of the investigation.

✷ Write these statements, filling in the gaps:
After the first experiment, hydrogen was found in ______ (tube H only/tube A only/both tubes). After the second experiment, hydrogen was found in ______. The hydrogen diffuses ______ (upwards/downwards/in all directions) through the air.

5.3 Diffusing speeds

Do different gases diffuse at the same speed or at different speeds? You could find out with an apparatus like that in Fig. 5.4.

Investigation 5.2 Diffusion speeds

You will need:
the apparatus shown in Fig. 5.4

As soon as the bungs and wads of cotton wool are put into the tube,

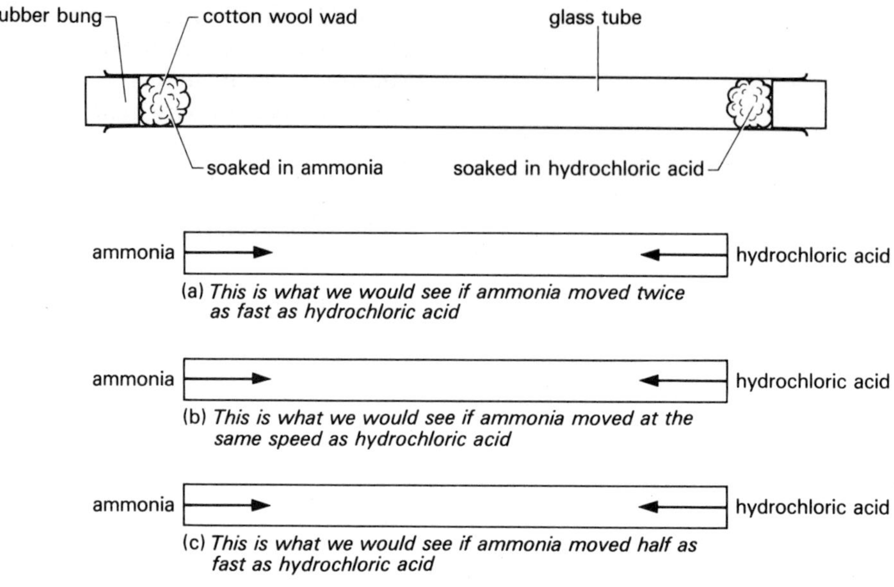

Fig. 5.4 Investigation 5.2

ammonia gas will start diffusing from one end and hydrogen chloride gas from the other. When they meet they will produce a disc of white smoke (a substance called ammonium chloride).

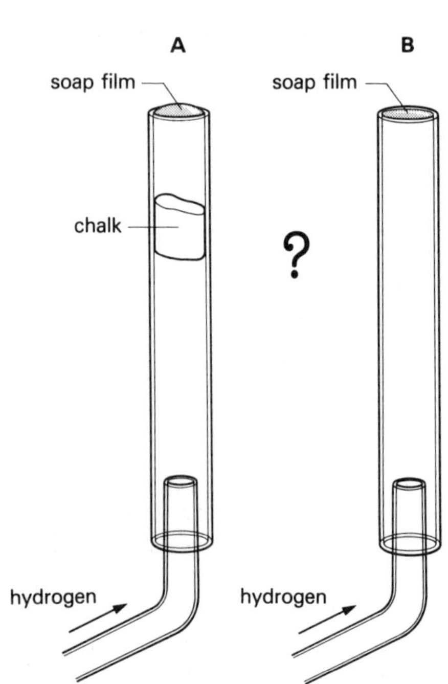

Fig. 5.5 Why does a soap bubble blow on the top of tube A but not on the top of tube B?

IN YOUR NOTEBOOK

✳ Write a heading: *Diffusion speeds*

✳ Copy the diagram of the apparatus.

✳ Copy the three other diagrams and the statements with them.

✳ Predict where the discs of white smoke will form in diagrams (a), (b) and (c). Mark the positions of the discs on each diagram.

Try the experiment and watch where the disc of smoke forms.

Fig. 5.5 shows a puzzle experiment to do with diffusion. Soap solution is smeared on the tops of both tubes. A soap bubble blows on the top of tube A but not on B. Can you explain why?

Diffusion was a mystery to scientists for hundreds of years. They knew all about it but could not explain how and why it happened.

The next section tells about another scientific mystery. Later on you will find that both mysteries are explained by the same modern theory.

5.4 A mysterious jiggling!

One day about 150 years ago, a Scottish scientist called Robert Brown* was looking at some pollen grains under a microscope. He saw that the grains were jiggling about in an erratic way. He checked that the air around the pollen grains was quite still and that his microscope was all right.

At first he thought that the pollen grains might be alive, so he replaced them with tiny bits of dye floating in water. The bits of dye jiggled too.

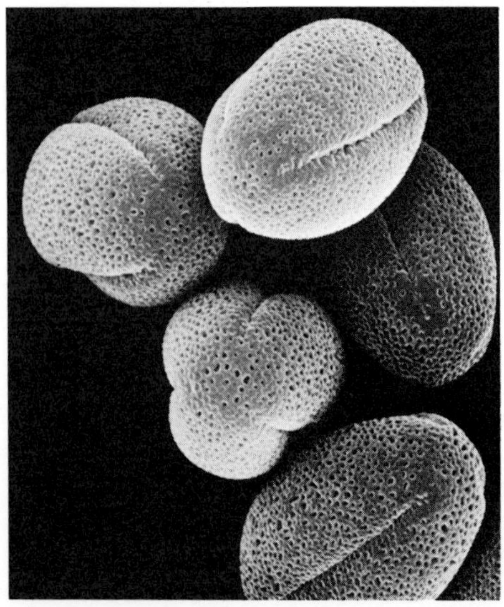

Fig. 5.6 Pollen grains magnified 1000 times

* Robert Brown is famous for another discovery. He was the first person to notice that most cells have a nucleus (see 'a very important blob' on page 76). He invented the word nucleus from a Latin word meaning 'little nut'.

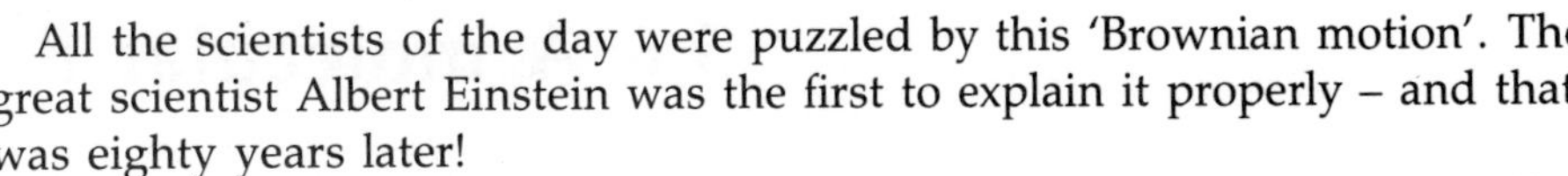

All the scientists of the day were puzzled by this 'Brownian motion'. The great scientist Albert Einstein was the first to explain it properly – and that was eighty years later!

If you have the right apparatus, you can see Brownian motion yourself.

Investigation 5.3 Looking at Brownian motion

You will need:
a 'smoke-cell apparatus'
a power supply
a microscope with a low power objective ($\times 7$ will do)

Fill the smoke-cell with smoke as in Fig. 5.7 and cover it. Focus the microscope (wind it down before you look through it, until it nearly touches the smoke cell, and then slowly back as you look through it).

IN YOUR NOTEBOOK

✱ Write a heading: *Brownian motion*

✱ Draw a diagram of the apparatus.

✱ Write these statements, filling in the gaps:
Under a microscope, smoke particles can be seen to jiggle about in an _______ way. This is called Brownian motion. My explanation for the jiggling about is _______.
(At this stage, your explanation is as good as anyone else's!)

5.5 A theory of gases

Diffusion and Brownian motion were partly explained by the Swiss mathematician Daniel Bernoulli in 1738. His theory is quite simple but it needs a good imagination!

Think of a balloon full of gas – full of air or hydrogen gas for example. The idea behind Bernoulli's theory is that the gas in the balloon is made up of millions of tiny particles which we now call *molecules*. There are millions of these molecules whizzing about, bumping into each other and into the walls of the balloon.

When you feel the pressure of a gas, you are feeling the results of 'bombardment' by millions upon millions of fast-moving molecules.

Molecules are too small to see – even through the most powerful microscope. They are so small that you could line up 100 000 across a full stop on this page.

Bernoulli's idea was so incredible (invisible particles, zipping about, indeed!) that no one took it seriously at the time. It was not fully accepted until a hundred years later.

Nowadays, Bernoulli's theory is called *the kinetic theory of gases*. The word 'kinetic' comes from the same Greek word as 'cinema' ('kinema' in the old days – a place showing moving pictures). It means 'concerned with moving things'. The theory has been fully worked out and involves some complicated maths. It can be used to explain many scientific observations. For instance, Einstein used it to explain Brownian motion. His idea was that the pollen grains jiggle about because they are being bombarded by invisible air molecules. Every so often, a pollen grain is hit by lots of air molecules going in the same direction. They make it jerk. Pretty unlikely, you might think, but there are so many molecules zipping about that such accidents can happen. Einstein worked out the chances of it happening to a particular pollen grain.

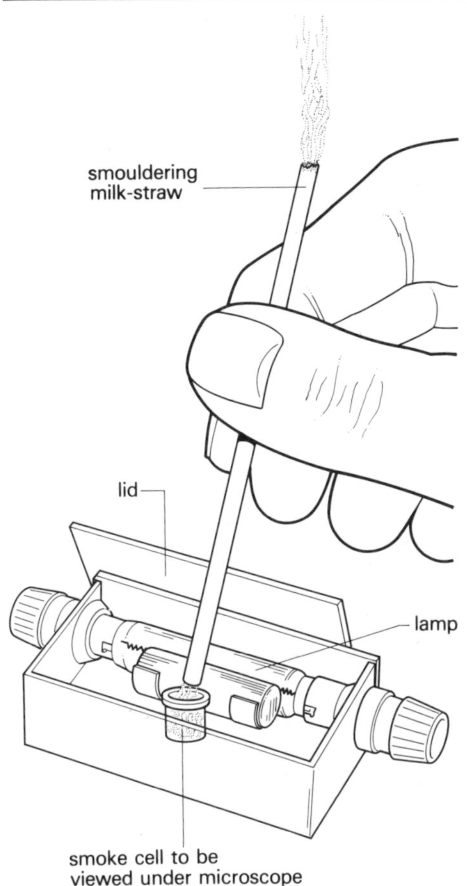

Fig. 5.7 Smoke cell apparatus

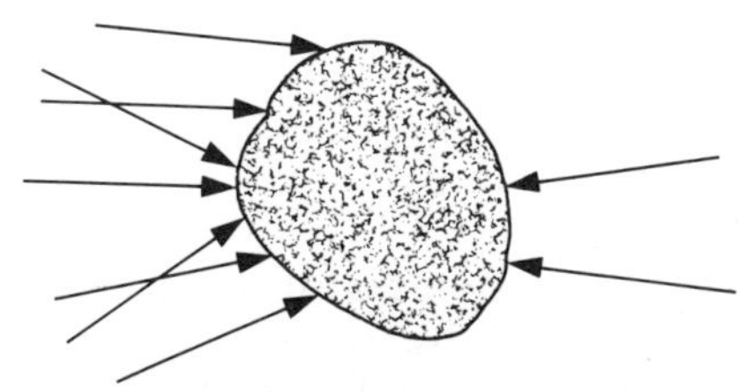

Fig. 5.8 The 'pollen grain' in the picture is shown a thousand times larger than life. The molecules, which are 20 000 times smaller, are far too small to see.

5.6 A molecular flick-book

You can get a moving picture of gas molecules moving about in a container if you make a 'flick-book'.

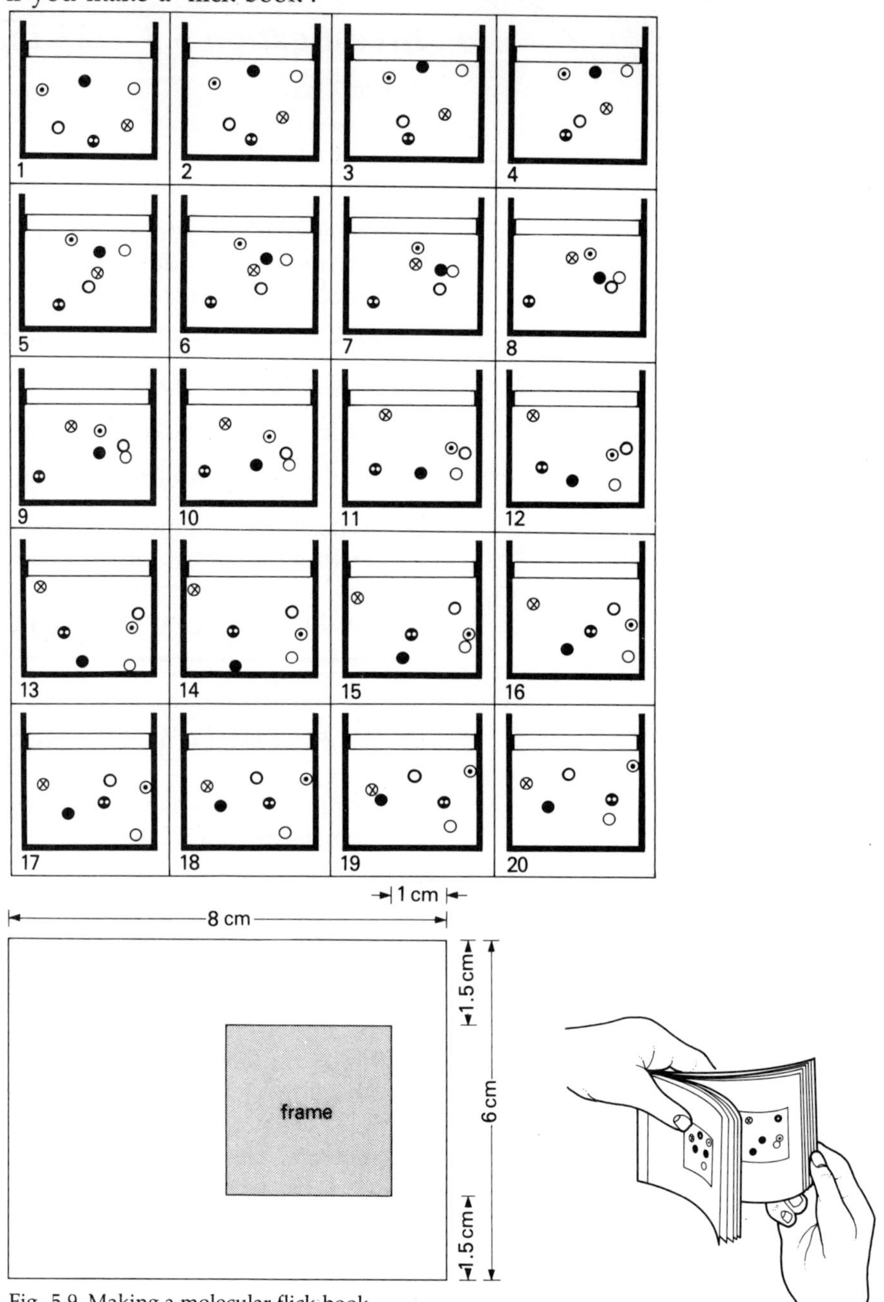

Fig. 5.9 Making a molecular flick-book

You will need:

20 pieces of thin card, each 8 cm × 6 cm   scissors, pens, pencils
tracing paper   the 20 pictures in Fig. 5.9

1 Carefully mark up the pieces of card as shown.
2 Trace one of the pictures on to each piece of card so that it just fits into the rectangle marked 'frame'.
3 Arrange the 20 pieces of card in order so that number 1 is at the bottom. Staple them together down the left hand edge.
4 Curl back the cards and let them flick down one by one. Watch one molecule at a time – the black one for instance.

5.7 The random walk game

This is another way to get an idea of molecular motion.

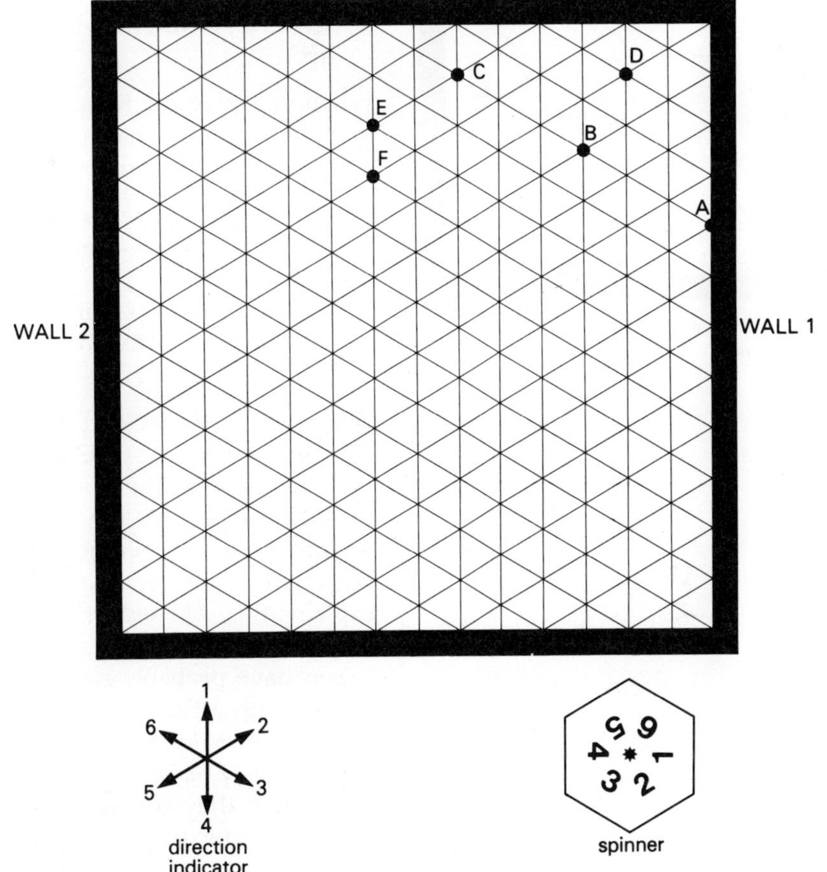

Fig. 5.10 Things needed for the 'random walk game'

You will need:

A playing board like the one in Fig. 5.10. (You can make your own out of a
sheet of so-called 'isometric' graph paper.)

A spinner made by tracing the diagram on to a piece of thick card, and
pushing a 'cocktail stick' or a pointed matchstick through the middle.

Two small markers – small beads or map pins for example.

Rules

✳ The game is for two to four players.

✳ Each players starts with his marker at the edge of one wall.

✳ Each player in turn spins the spinner twice. The first number tells him how
many spaces to move the marker. The second number tells in which
direction. For example, suppose the first player gets 3 and 6 for the two
spins. He would move his marker from point A to point B, 3 spaces away.

✳ *Reflection*
If the marker reaches a wall as part of its move, it is 'reflected' back to use
up the number of spaces in the move. For example, suppose the marker
was at point C and the spins were 4 and 2. The move would be from C to

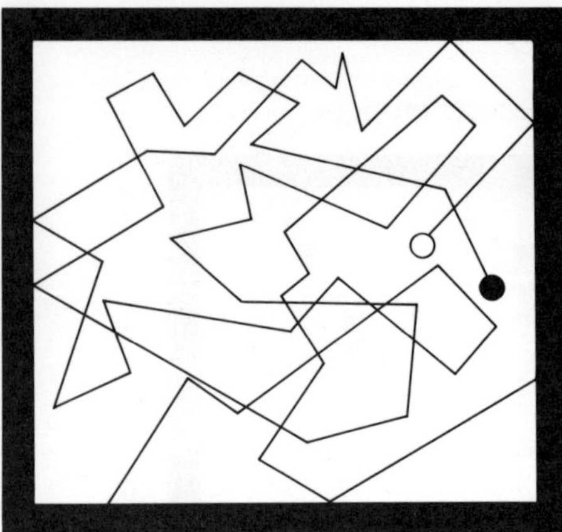

Fig. 5.11 Model of the way that molecules move in a gas

D, bouncing off the wall on the way. If the marker was at E and the spins were 5 and 1, the move would be from E to the wall and off again to F.

✻ *Scoring*
Every time a player's marker hits another player's wall, the first player scores a point. The winner is the player with most points when the time is up.

When you play the Random Walk Game you are making a slow-motion model of the way that molecules move in a gas. Your markers stand for molecules moving around, colliding with the walls of their container.

1 What would happen to the number of collisions if you halved the size of the playing board – moving the 'walls' inwards? Would you get more in a game – or less?
2 What would happen if you moved your markers faster? What if you doubled the number of spaces covered by each move? Would you get more collisions or less?

When you have decided on your answers, compare them with your friends'. Do they agree with you?

Have you ever put your finger over the nozzle of a bicycle pump and pressed the plunger down? It makes an interesting noise! If you make the air go into a smaller volume, its pressure increases until it rushes out past your fingers. The answer to question 1 above tells you how the kinetic theory explains this: *The smaller the volume, the more often the molecules collide with the wall, so the more they press.*

You have probably seen the experiment in which a sealed can of air is heated with a Bunsen flame. How does the kinetic theory explain the result? The answer to question 2 helps. The explanation is that *the higher the temperature of the air, the faster its molecules move.* The faster they move, the more often they collide with the walls, so the more they press. On top of that, the faster they move, the harder they 'bang' against the walls!

IN YOUR NOTEBOOK

✻ Write a heading: *Gas patterns and kinetic theory explanations*

✻ Copy this table, filling in the gaps:

Gas pattern	Explanation
Gases spread out to fill any space. They move through and mix with other gases. This is called ______.	Gases are made up of tiny particles called ______. These are moving about all the time.
The smaller the volume of a gas, the ______ the pressure it exerts. (Call this the PV pattern.)	Decreasing the volume ______ the number of collisions between the molecules and the walls of the container.
The higher the temperature of a gas the ______ the pressure it exerts. (Call this the PT pattern.)	Heating a container of a gas increases the speed of the gas molecules.

5.8 Gas patterns in action

About a century ago, a German inventor called Nikolaus Otto built the first version of today's motor car engine. That is, the first *internal combustion* engine. Otto used the gas patterns – the *gas laws* – to work out his design for the engine.

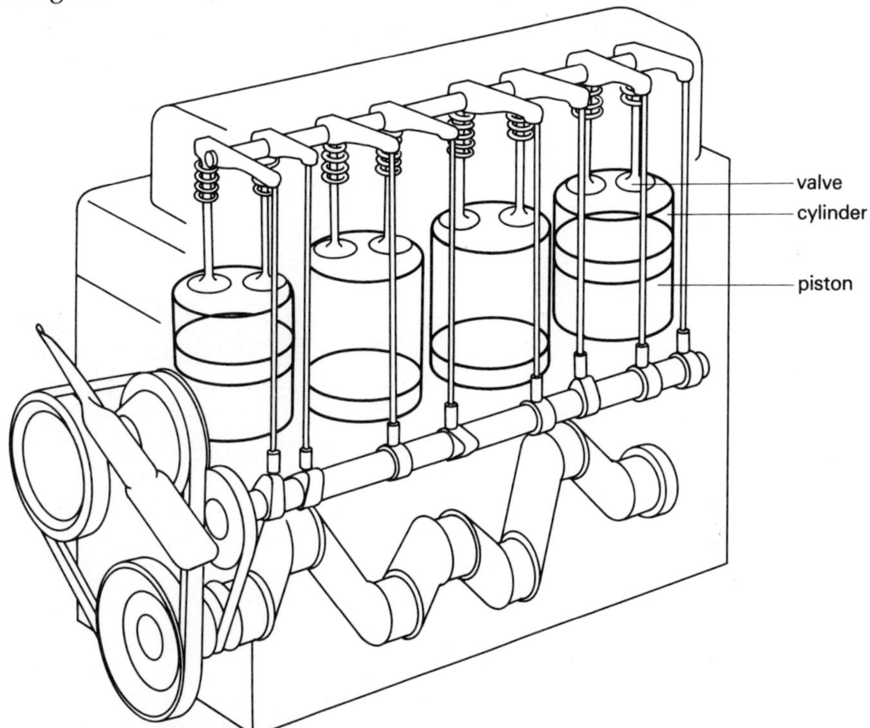

Fig. 5.12 Section through a typical modern engine

Fig. 5.12 shows a picture of a typical engine. Lined up along the middle of the engine you can see four fat tubes, called *cylinders*. Each has a *piston* sliding up and down inside it. As the pistons slide up and down they turn the oddly-shaped shaft at the bottom, through linking rods which are not shown in the diagram.

Fig. 5.13 shows what happens inside *just one* of the cylinders when the engine is running. It takes less than a fiftieth of a second for all five stages.

1 The piston is moving downwards. The volume in the cylinder is getting larger, so the pressure in it is tending to drop (the PV pattern). Petrol and air are coming in to make up the difference.
2 The piston is moving upwards, squashing the mixture of air and petrol. The volume is getting smaller so the pressure is going up (the PV pattern again).
3 Just before the piston reaches the top, a special 'spark plug' sets fire to the petrol–air mixture. Now there is an explosion and the temperature shoots up. So the pressure shoots up as well (the PT pattern).
4 The piston is moving downwards pushed by the exploding gases. The volume is increasing so the pressure is falling.
5 The piston is moving upwards, the volume is getting smaller and the pressure is tending to rise. The result is that the 'exhaust' gases are pushed out.

The piston has to travel the length of the cylinder four times to get back to the beginning again; so this kind of engine is called a *four-stroke* engine.

Look back at the complete engine. Notice that the movement of the

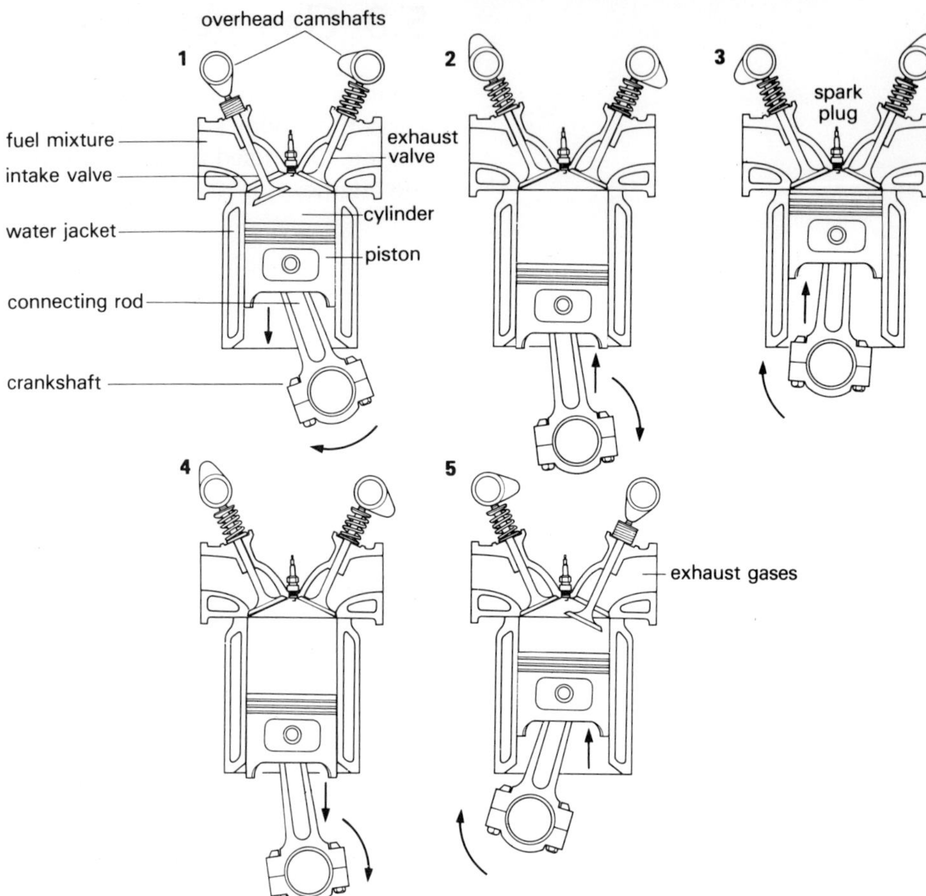

Fig. 5.13 The series of events inside *one* of the cylinders when the engine is running

pistons is staggered so that there is always one being pushed down by an explosion.

SOME THINGS TO FIND OUT

1 What is a two-stroke engine?
2 How does a diesel engine work?
3 What is a Wankel engine?
4 What does a carburettor do?

Write about these things in your notebook.

5.9 Solids and liquids

What about solids and liquids? Are they made up of molecules? If they are, do the molecules move or not?

As with gases, we cannot see the molecules so we cannot see if they are moving. But we have to suppose that they are there and that they do move. It is the only way we can explain the results of experiments like those in the next investigation.

Investigation 5.4 Solid–solid and liquid–liquid diffusion

You will need:
a Petri dish containing some hard set gelatine   narrow cork borer
a plastic syringe held on a stand   a small container of water

a tiny crystal of potassium dichromate,
malachite green or congo red

a small container of ink

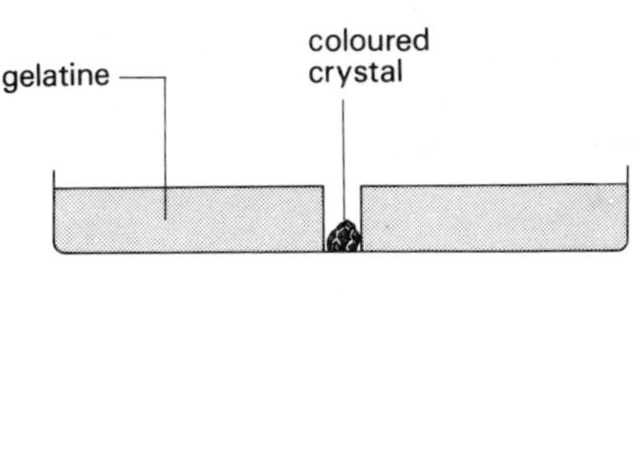

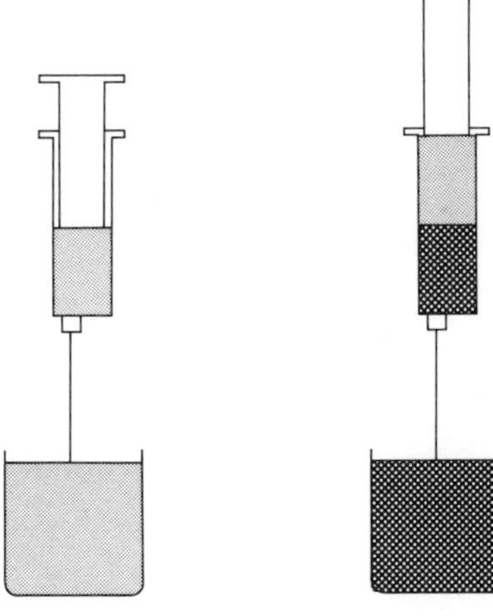

Fig. 5.14 Investigation 5.4

1 Make a well in the gelatine with the cork borer.
2 Put the crystal in the well. Make sure that it touches the gelatine.
3 Look at it again after a few days.
4 Arrange the syringe and stand so that the needle dips in the container of water. Draw in half a syringe full.
5 Now arrange the syringe and stand so that the needle dips in the container of ink. Very slowly and carefully, draw ink into the syringe as shown in Fig. 5.14.

IN YOUR NOTEBOOK

✳ Write a heading: *Moving molecules*

✳ Write these statements, filling in the blank:

Most of the things that gases do can be explained by imagining that they are made up of tiny particles called _______. These particles are extremely small and move about very fast bumping into each other and into the walls.

The higher the temperature of the gas the faster the molecules move.

The molecules of solids and liquids also move, but in different ways to those of a gas.

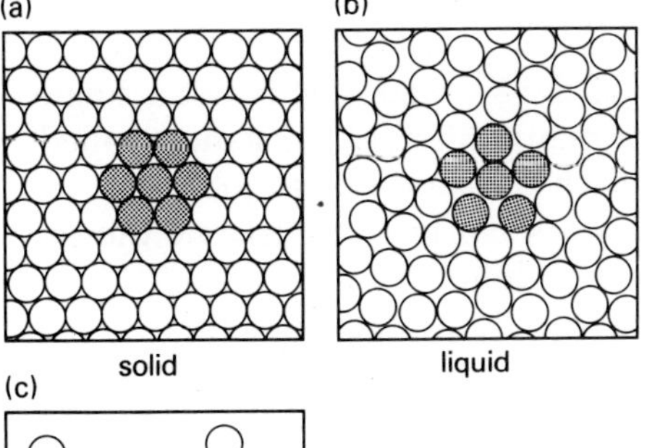

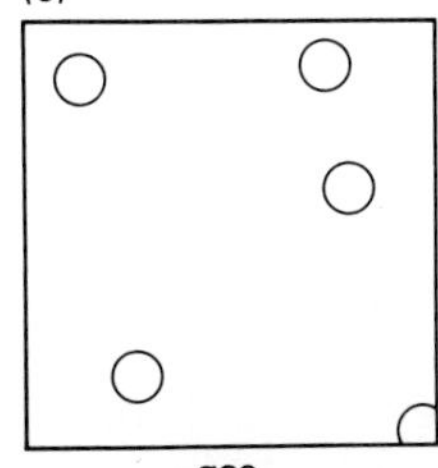

Fig. 5.15 The arrangements of molecules in solids, liquids and gases

5.10 Solids, liquids and gases

Suppose you heated some silver metal until it reached 960 °C and melted. What would be happening to its molecules? Suppose you went on heating until the silver boiled at 1950 °C. (You would need a special furnace.) What would happen to its molecules then? What does the molecular theory say about melting and boiling?

The theory says that two things happen when a substance melts – when it changes its *state* from a *solid* to a *liquid*. The arrangement of the molecules changes and there is a change in the way they move. The temperature when this happens is called the *melting point* of the solid.

In a solid, the molecules are arranged in a regular way – perhaps as in Fig. 5.15(a). In a liquid the arrangement is much less regular.

When a substance *vaporises* – changes from a *liquid* to a *gas* – the molecules move apart and go off on their own. The temperature when this happens to all parts of the liquid at the same time is called the *boiling point*.

In a solid, the molecules stay more or less where they are, jiggling about in a small space. You can see what happens in a liquid. As well as jiggling about, the molecules move bodily from place to place.

One other thing you should know about molecules is that some are more massive than others. The following pages use this idea to explain an important industrial process.

5.11 Dividing up the fractions

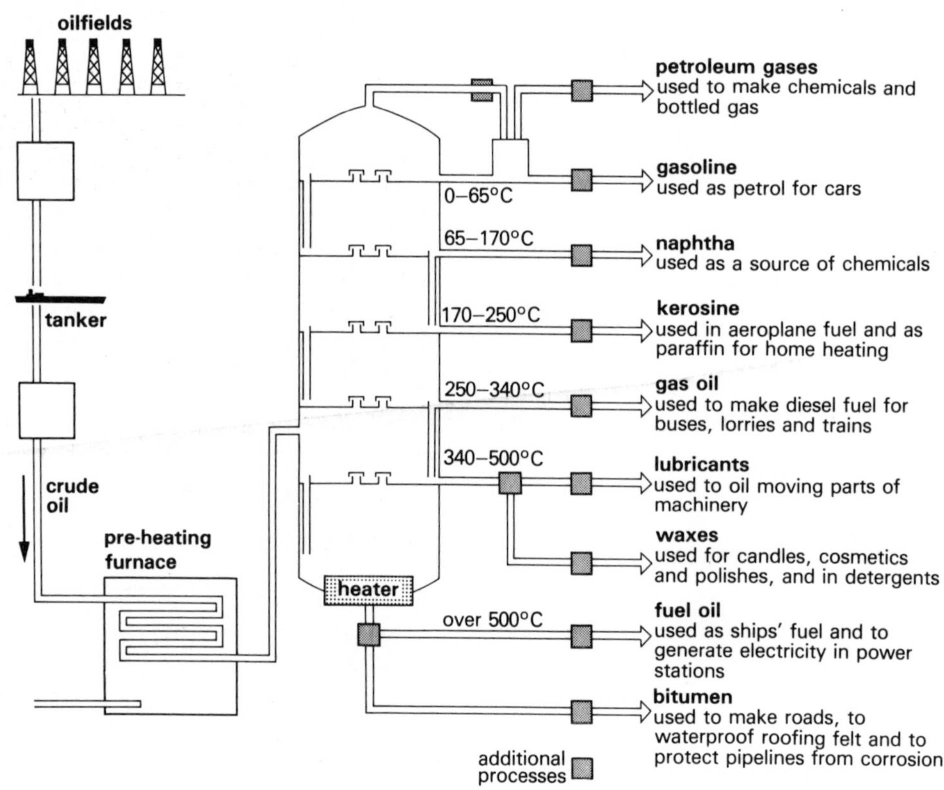

Fig. 5.16 Fractional distillation in industry

As it comes out of the ground – or out of the sea bed – oil is not a single substance. It is a complicated mixture of many different substances called *fractions*. Each fraction is useful to us – each does a different job. So scientists and technologists have had to work out how to separate the different fractions. Fig. 5.16 shows how it is done in industry.

You can use the kinetic theory of molecules to explain how the 'fractionating column' works. From page 91 you will remember that: *the higher the temperature the faster molecules move*. You will need a new piece of information: *The more massive a molecule is, the slower it moves at a certain temperature*.

Suppose oil was a mixture of only two substances – one substance with very heavy (very massive) molecules, and one substance with very light molecules. If you began to raise the temperature of the oil, the light molecules would speed up faster than the heavy ones. They would soon travel fast enough to fly off as a gas – leaving the heavy ones behind. So you could separate the two kinds.

Of course, oil is a mixture of many substances. If you raise the

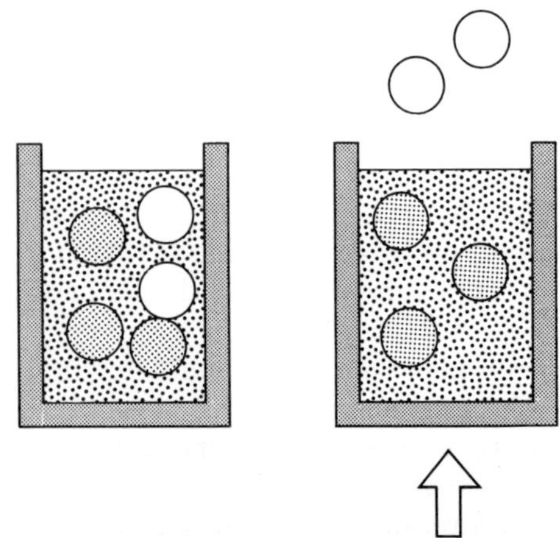

Fig. 5.17 Light molecules speed up faster than heavy ones when a mixture is heated

temperature steadily, those with lighter molecules will leave the oil first. Then the next heavier ones, then the next heavier, and so on until you are left with only the heaviest. Over a period of time the lighter, faster molecules will also go further than the heavier ones.

So you can see that, in the fractionating column, the lightest molecules will be found up at the top and the heaviest down at the bottom.

The top of the column is cooler so the gas will condense back into a liquid – and be piped away.

Question: Which substance has the heaviest (most massive) molecules: Gasoline (petrol) or Kerosine (paraffin)? Can you say why?

You can try this 'fractional distillation' for yourself. But you will need a science lab and *you should do the investigation in a fume cupboard.*

Fig. 5.18 Fractionating columns of an oil refinery

Investigation 5.5 Trying out fractional distillation

You will need:
a boiling tube with side arm
a little 'Rocksil'
a delivery tube with a rubber
  connector
a rubber bung fitted with a thermometer

a small beaker
four small test-tubes in a rack
a Bunsen burner
some crude oil from your teacher

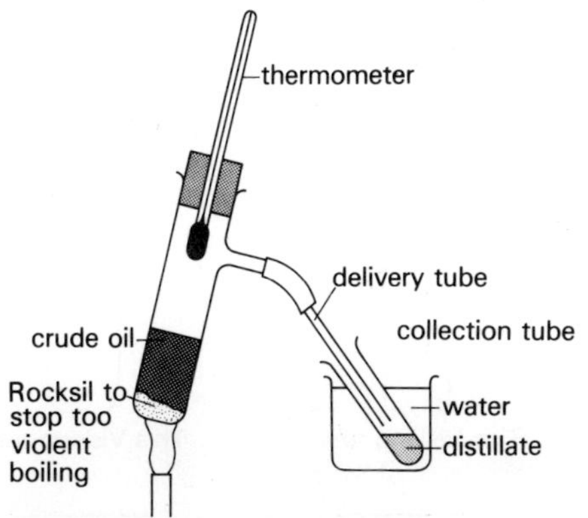

Fig. 5.19 Investigation 5.5

WARNING! *Some kinds of crude oil can be dangerous. You must only use the kind supplied by your teacher.*

1 Set up the apparatus. Warm the oil gently using a small Bunsen flame. Hold the burner by its base and waggle it around the tube.
2 Watch the thermometer and temporarily stop heating when the temperature reaches 65 °C. By alternately heating and cooling, try to keep the temperature around 70 °C for a minute or two.

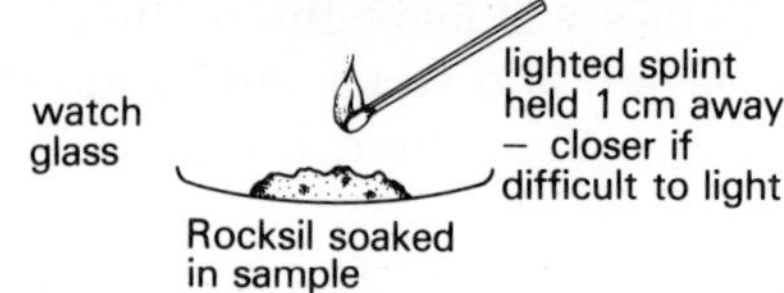

Fig. 5.20 The flammability test

3 When you have collected a small quantity of distillate, put the delivery tube into a new collection tube. Put the first collection tube back in the rack.

4 Heat the oil until the temperature is around 120 °C. Try to keep it there for a few minutes. Collect the distillate.

5 Repeat steps 3 and 4 for temperatures of 170 °C and 220 °C.

You should have four samples of oil fractions. Figure 5.20 shows one way of testing them. Which fraction is the easiest to light? Which is the hardest? Are there any other obvious differences between the fractions?

IN YOUR NOTEBOOK

✱ Write a heading: *Separating the fractions of crude oil*

✱ Draw a diagram of the apparatus.

✱ Write a few notes to say what you did.

✱ How easy was it to keep the temperatures steady?

✱ What differences were there between the fractions?

5.12 A molecular sieve?

You already know that molecules are constantly on the move. You know that the higher the temperature, the faster they move. And you know that some are more massive than others.

This investigation gives clues to another feature of molecules.

Investigation 5.6 The action of Visking tubing

You will need:
two boiling tubes
two bags made out of Visking tubing (a material like cellophane), open at one end and tied in a tight knot at the other
two paper clips
a strong* solution of cane sugar (sucrose)
a weak* solution of cane sugar

Set up the apparatus as shown in Fig. 5.21. Leave it for 45 minutes – or overnight if you cannot wait around – and look at it again.

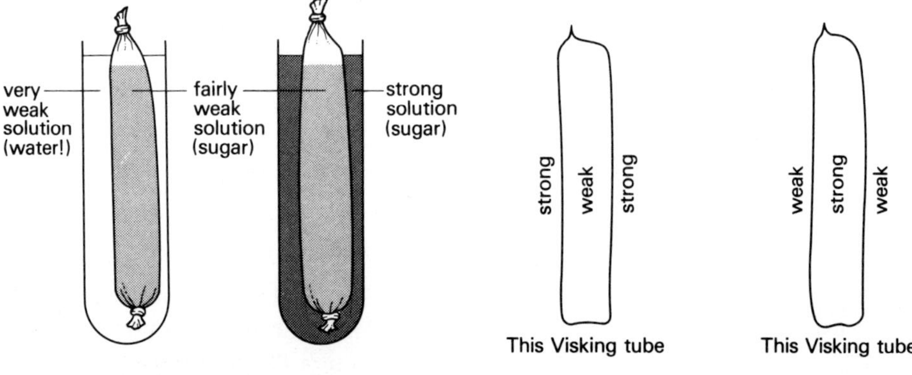

* The words 'strong' and 'weak' are used here in an everyday way. Scientists sometimes use them in a special way. You will read more about this later.

Fig. 5.21 Investigation 5.6

IN YOUR NOTEBOOK

✳ Write a heading: *An investigation with Visking tubing*

✳ Show the results of your experiment on diagrams like those in Fig. 5.21.

✳ Suppose a Visking tube shrinks if it loses water and swells if it gains water. Put arrows on your diagrams to show the way that water moves into or out of the two tubes.

✳ Write these statements, putting the words 'strong' or 'weak' into the correct gaps:
Water moves through the Visking tubing from a _______ solution into a _______ solution.

5.13 A molecular explanation

You can explain what happens by saying that *some molecules are bigger than others*. The Visking tubing acts like a sieve for molecules. It lets smaller ones through and stops larger ones. It has many small gaps (pores) which let the water molecules – which are moving of course – pass through. The gaps are too small to let the sugar molecules through. The material is called a *differentially-permeable membrane DPM* – meaning a thin sheet (membrane) which will allow some things to pass through but not others.

To start with, there are fewer water molecules on the sugar side because space is taken up by the sugar. Water moves in to put this right.

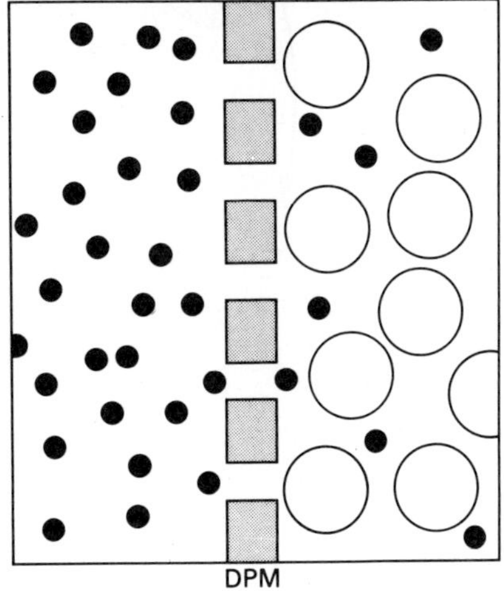

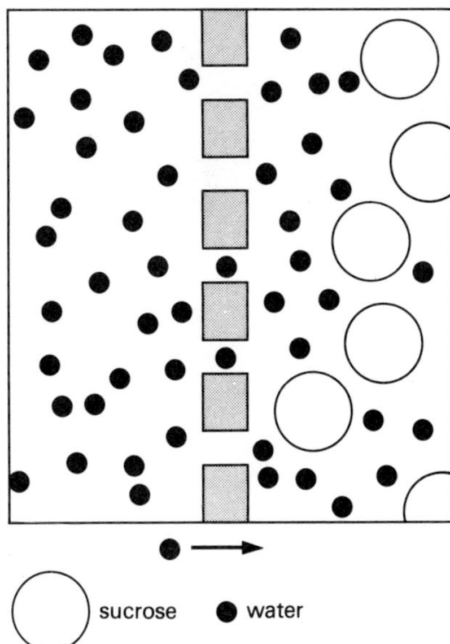

Fig. 5.22 Water molecules can pass through the DPM but sugar molecules cannot

IN YOUR NOTEBOOK

✳ Write this important statement:
The movement of water from a weak solution to a strong solution – through a differentially-permeable membrane – is called osmosis. *Using a differentially-permeable membrane to separate substances is called* dialysis. *Osmosis can be explained by imagining that some molecules are larger than others.*

The membrane of a cell is a differentially-permeable membrane. You can investigate the way it works with onion or rhubarb cells.

Investigation 5.7 Looking at cells in different solutions

You will need:

a red onion or a stick of rhubarb
a microscope
one or two microscope slides and
  cover slips
a scalpel
a dropper pipette

a small paint brush
forceps
a few strips of filter paper
a small amount of strong cane sugar
  (sucrose) solution in a beaker

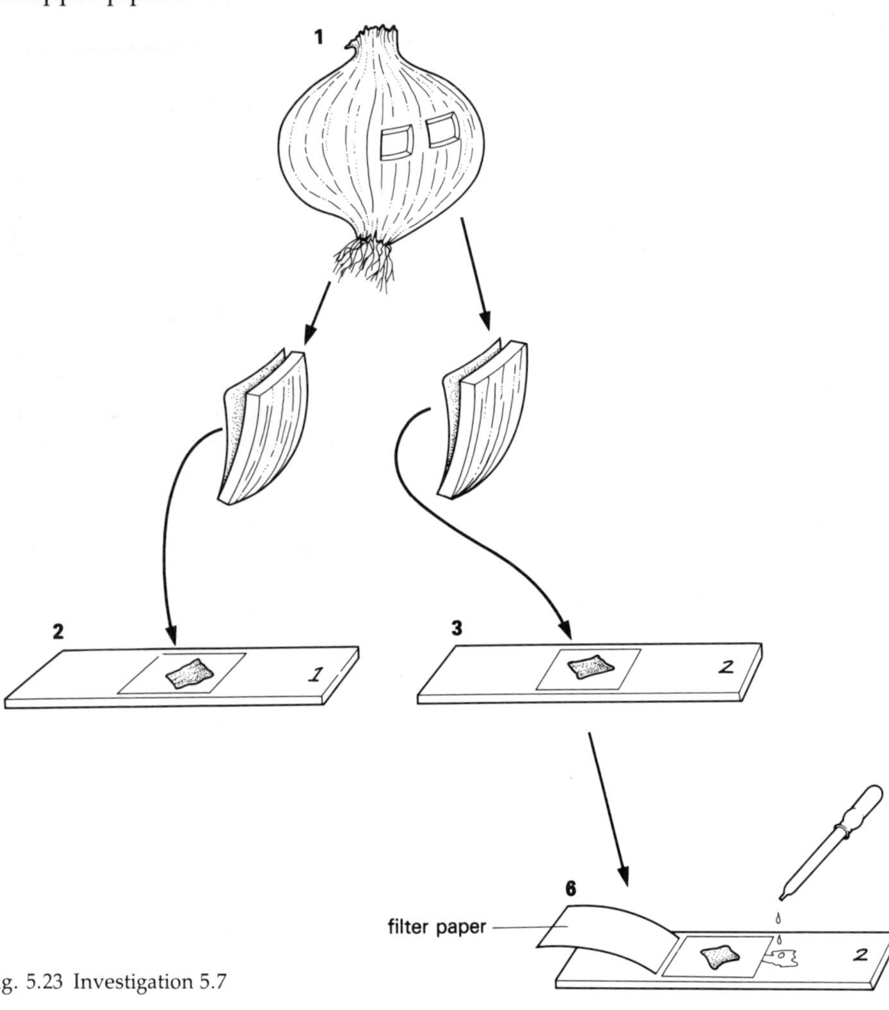
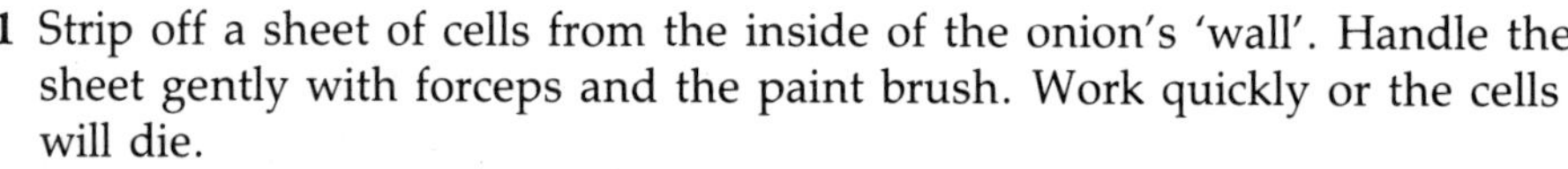

Fig. 5.23 Investigation 5.7

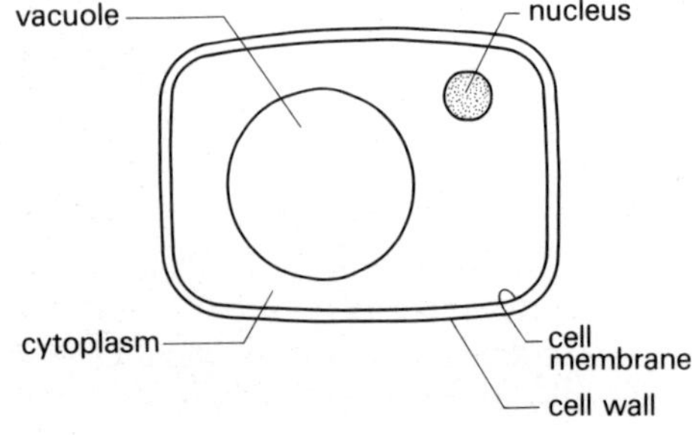

Fig. 5.24 A very simple diagram of a plant cell

1 Strip off a sheet of cells from the inside of the onion's 'wall'. Handle the sheet gently with forceps and the paint brush. Work quickly or the cells will die.
2 Put one sheet on a slide, add a drop of water and cover. (Slide 1)
3 Mount another sheet in sugar solution. (Slide 2)
4 Watch the cells under a microscope for a few minutes.
5 Draw sketches to show any changes in the cells. Copy Fig. 5.24 then draw it twice more to show how it looks in water and in sugar solution.
6 After a few minutes, wash the sugar solution from slide 2 using a dropper pipette and a filter paper to draw the water through.
7 Look at slide 2 again under the microscope. Sketch the cells.

IN YOUR NOTEBOOK

✳ Write a heading: *Cells in different solutions*

Copy your cell diagrams and label them: *'in water'* and *'in sugar solution'*.

✳ Write these statements, filling in the gaps:

In tap water (slide 1), the onion cells seem to ______.

In sugar solution (slide 1) the cells seem to ______.

Water moves into the onion cells in slide ______ *and out of them in slide* ______

O ______ *is happening. The 'skin' of the cells is a* ______ - ______ *membrane.*

5.14 DPMs in action

One way of preserving fruit is to surround it with thick sugar syrup (a strong solution of sucrose). The syrup kills any bacteria that get into it. Can you say how?

To answer the question, you need to know that most bacteria are single cells, and that their cell membranes are differentially-permeable. Also, that the solutions they contain are much weaker than the sugar syrup.

This is a much-magnified picture of a plant's root. You can see the tiny 'root

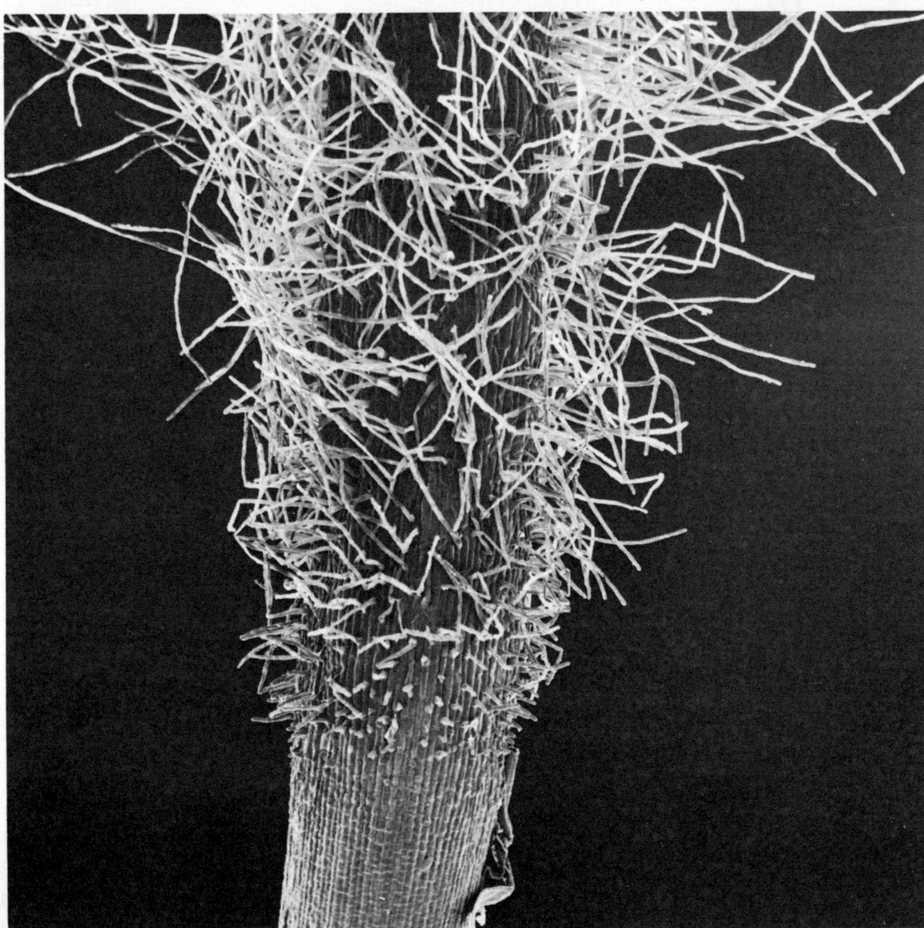

Fig. 5.25 A root magnified 150 times to show the root hairs

hairs' which grow all over the surface of the root except at the tip. The hairs, which are parts of special cells, do an important job for the plant. They collect water from the soil. Why does the water go into them? Why doesn't it come out again?

To answer, you need to know that the root hair cells contain a strong solution of various substances. The root hairs are bathed in soil water. Their walls are differentially-permeable membranes.

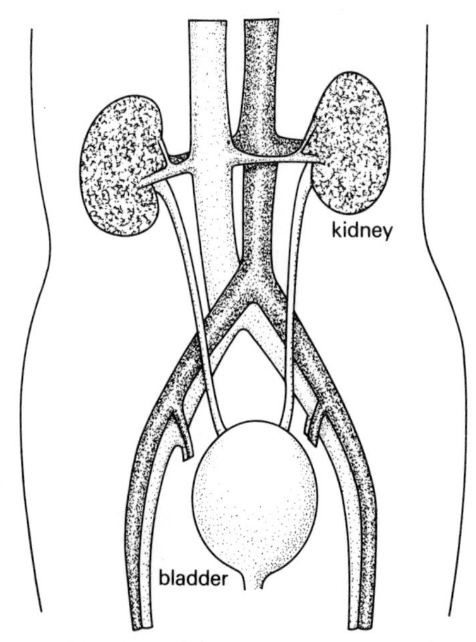

Fig. 5.26 The position of the kidneys in the body

5.15 Molecules and the kidney machine

Every year, thousands of people die because their kidneys stop working properly. A number of diseases can cause this to happen. The lives of a lucky few are saved by connecting them to kidney machines.

You can use what you know about the kinetic theory of molecules and osmosis to explain how a kidney machine works.

Like other animals, humans need a daily intake of protein. Unwanted protein cannot be stored and is broken down into a poisonous substance called *urea*. Normally, our kidneys quickly remove the urea, together with other waste products, and we excrete them in urine. As well as cleaning the blood, the kidneys do an important balancing job. They keep the amount of water, blood sugar, and salts (such as sodium chloride) at a steady level by getting rid of any excess. If the kidneys fail, poisons and unwanted substances soon build up to dangerous levels.

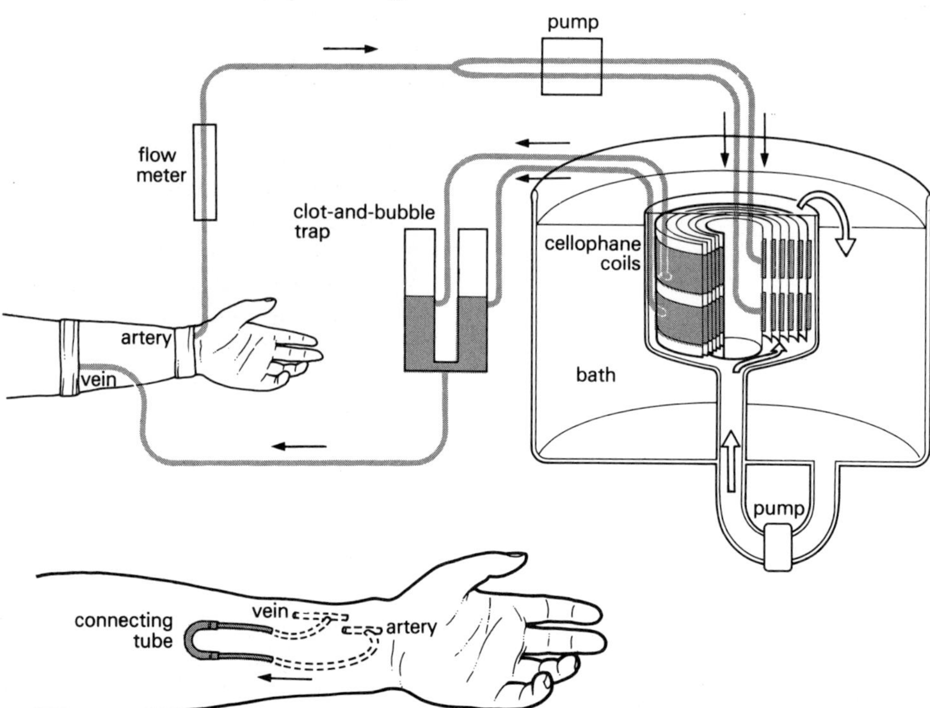

Fig. 5.27 The kidney machine. The diagram of the arm shows the plastic tubes connected together when the patient is not using the machine. The tubes remain in the arm all the time.

The kidney machine takes over the job of the kidneys. The patient's blood is passed through flat cellophane tubes coiled round and round in spirals. The spirals are bathed in a solution of salt and blood sugar (glucose) in water. Urea and other poisonous substances pass out of the blood through the cellophane. The blood cells and valuable proteins stay in the blood. The water, salt, and sugar pass in both directions, 'balancing up' the blood.

The kidney machine was invented in German-occupied Holland during the Second World War. Its inventor, Willem Kolff kept it secret, and used it to save the lives of a number of resistance fighters. It worked very well but had one disadvantage. So that blood could be passed through the machine and returned to the patient, quite large tubes had to be pushed into a vein and an artery. This took difficult and painful surgery.

In 1960, Dr Belding Scribner in the USA worked out how to put plastic tubes into a fat vein and a large artery so that they could stay there all the time. They stick out of the patient's body and are connected together when the patient is not using the machine.

Checkout

Keywords

cell	membrane
compression	molecule
dialysis	organ
differentially-permeable	osmosis
membrane (DPM)	pressure
diffusion	random motion
expansion	temperature
fractional distillation	volume
kinetic theory	

Patterns

1 Gases tend to spread out to fill a space, perhaps moving through and mixing with other gases as they do so (diffusion).

2 A gas presses against the walls of its container. The pressure increases if: (a) the volume of the container is reduced; (b) the temperature of the container is raised.

3 Through a microscope, some solid particles can be seen to be constantly moving (Brownian motion).

4 If a strong solution and a weak solution are kept apart by a certain kind of membrane (DPM), the solvent (e.g. water) will move from the weak side to the strong side (osmosis).

5 If a mixture of different liquid substances has its temperature steadily raised, each substance in turn will leave the mixture as a gas (fractional distillation).

A theory

The kinetic theory can be used to explain diffusion, gas pressure, Brownian motion, osmosis, and fractional distillation. The theory says that:

(a) Solids, liquids, and gases, are made up of very tiny particles (molecules).

(b) The molecules are constantly moving in a random way.

(c) The molecules in a solid are vibrating. Those in a liquid are vibrating and moving bodily. Those in a gas are moving bodily.

(d) The higher the temperature, the faster the molecules move.

(e) Molecules of different substances can have different sizes and different masses.

(f) The greater the mass of a molecule, the slower it moves at a certain temperature.

True or false?

You will need a metal can with a screw top, and a supply of heat.

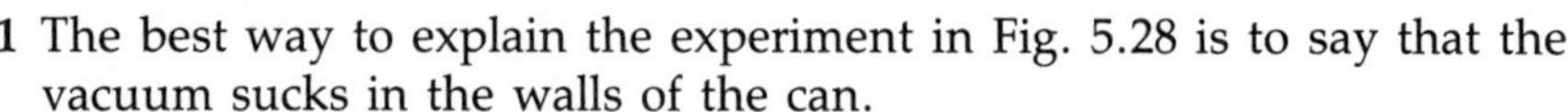

Wash the can out and fill it about one third full with water.

Make sure the lid is off. Heat the can over a flame until the water boils.

When steam is pouring out of the top, turn off the heat and *at the same time* screw the cap on tightly.

Run cold water over the can — **watch**.

Fig. 5.28 A vacuum experiment

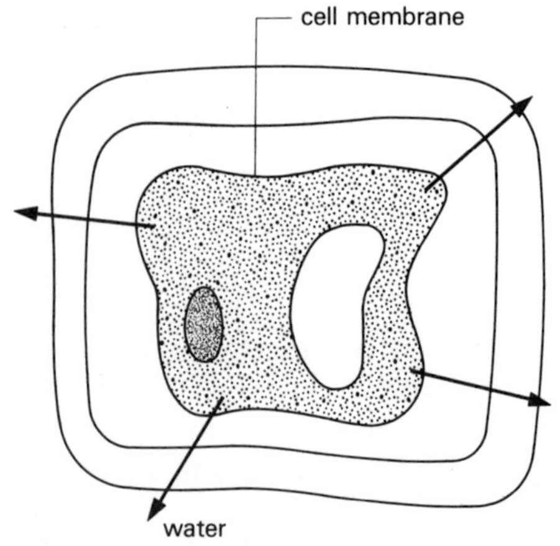

Fig. 5.29

1 The best way to explain the experiment in Fig. 5.28 is to say that the vacuum sucks in the walls of the can.
2 The higher the temperature, the faster a gas is likely to diffuse.
3 All gases diffuse at the same rate at the same temperature.
4 Pollen grains are roughly the same size as molecules.
5 Under a very powerful microscope you can see molecules moving about.
6 If you could take all of the molecules out of a container of oxygen gas you would be left with a perfectly empty space.
7 Fig. 5.29 shows what will happen if an onion cell is put into a bath of pure distilled water.
8 If a field near the coast was flooded by sea water, the root cells of plants in the field would lose water and shrink.
9 A differentially permeable membrane with allow any liquid to pass through it.
10 If you boiled a container of crude oil until only a few drops of liquid were left, the drops would be made up of the lightest fraction.

Problems

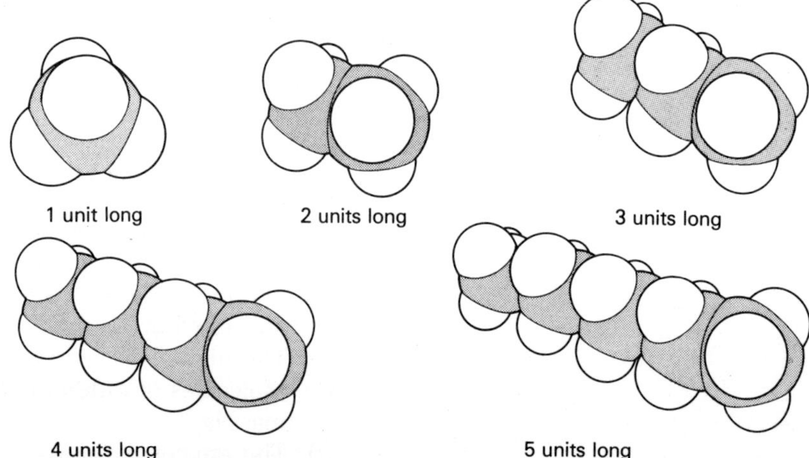

Fig. 5.30 Molecular models of some of the alkanes

1 There is a family of chemical substances called the 'alkanes'. They are made up of the elements carbon and hydrogen. Fig. 5.30 shows models of the molecules of five members of the family. The table shows their approximate melting points and boiling points.

Alkane	MP/ °C	BP/ °C
methane	−186	−164
ethane	−169	−103
propane	−187	− 45
butane	−135	1
pentane	−130	36

Draw a graph of the boiling points of the five alkanes against their lengths in 'units'. Try to answer these questions.

(a) The next alkane in the series is called hexane. It is six units long. Predict its boiling point.
(b) How many of the five alkanes in the table are liquid at 20 °C?
(c) According to the kinetic theory, which of the five alkane molecules would be moving fastest at 20 °C? Explain your answer.

2 Fig. 5.31 shows an apparatus which is used to study the way that the pressure of a gas changes when its volume is allowed to increase. The table shows a set of results obtained with the apparatus.

Pressure, P 1000 N/m^2	Volume, V cm^3*	PV
300	16.0	
280	17.5	
260	18.75	
240	20.25	
220	22.0	
200	24.25	
180	27.0	
160	30.25	
140	24.0	
120	40.0	

(a) Plot the results on a graph of Pressure against Volume. Draw a smooth curve through the points.
(b) Use the graph to find out what happened to the volume as the pressure changed from 120 000 to 240 000 N/m^2. Write this statement and complete it:
When the pressure was doubled the volume was ______.
(c) Use an electronic calculator to work out the P multiplied by V values which go into the third column. Is there a pattern to the PV values? Write it down in your own words.

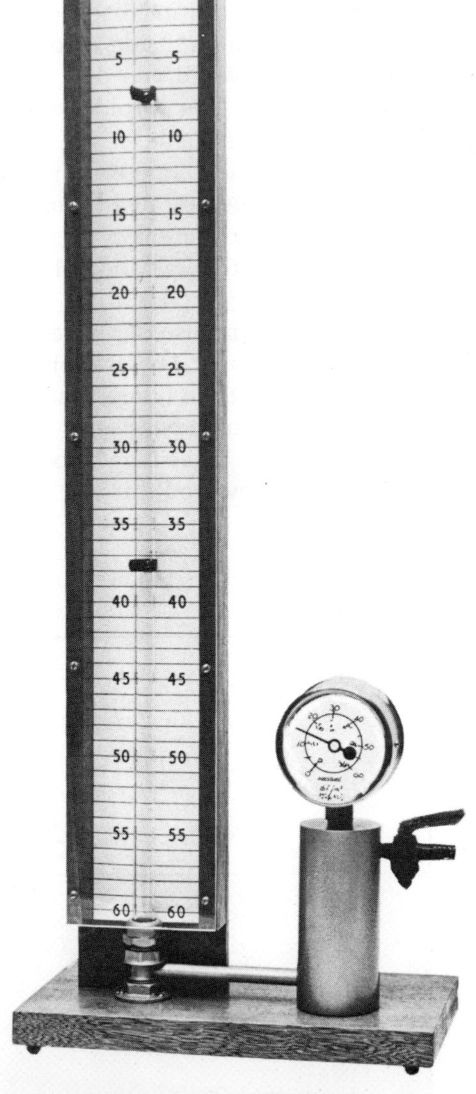

Fig. 5.31 Apparatus used to measure pressure changes in a gas

* m^3 would have been better units, but then the figures in this column would have been very small.

6 Atom

6.1 How old is the word atom?

Do you know how old the word *atom* is? Is it a modern word – less than 50 years old – or is it a bit older than that – a few hundred years old? What do you think?

The answer might surprise you. 'ATOM' is nearly 2500 years old! It comes from a Greek word meaning 'uncut' or 'smallest part'. Around 400 BC a few Greek scholars decided that everything was made up of tiny little bits of stuff too small to see. It was only a hunch – they could not prove it – but they gave us the word.

Between 400 BC and now, scientists have made complicated theories about atoms. The strange thing is that no-one has ever seen one. Atoms are so tiny that they are invisible – even through the most powerful microscope. Learning about them has been rather like finding out what is inside a box without opening the lid. If you had that problem, what would you do?

You would probably shake the box and listen carefully. You would use what you know about different sounds to guess what was inside.

This chapter is partly about 'shaking the atomic black boxes'. It looks at some of the evidence that scientists have used to make their guesses about atoms. And it looks at some useful things that have happened on the way.

How can you get an idea of the size of an atom? Figs. 6.2, 6.3 and 6.4 might help.

Fig. 6.2 shows a mosquito – the kind that carries malaria disease. It has been magnified ten times. It is looking at a tiny grain of salt. The grain is actually about the size of a full stop on this page. Look at the little square around the outside of the salt grain. In Fig. 6.3 the little square has been magnified ten times. Everything else in the picture is ten times bigger too. The salt grain and the mosquito's limbs look the way they would if you had

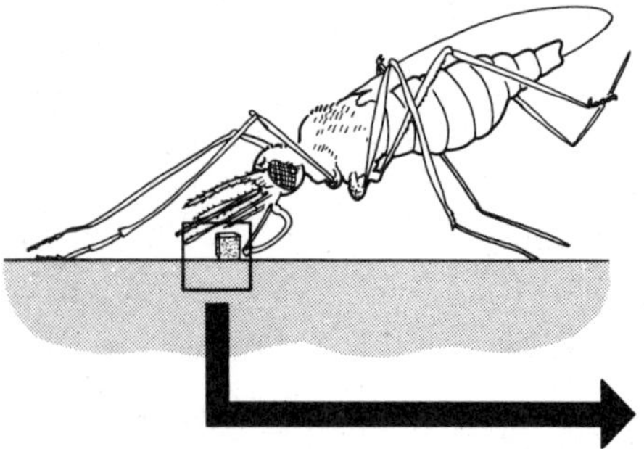

Fig. 6.1 The word 'atom' is nearly 2500 years old

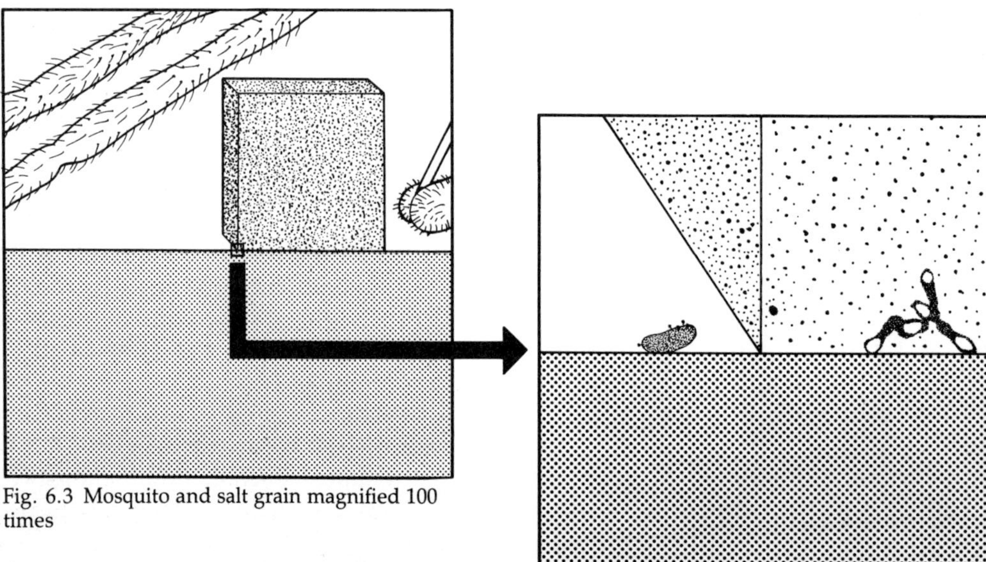

Fig. 6.2 Mosquito and salt grain magnified 10 times.

Fig. 6.3 Mosquito and salt grain magnified 100 times

Fig. 6.4 Salt grain magnified 10 000 times

shrunk to a hundredth of your normal size. For instance, if you are about 1.6 metres tall, you would need to be 1.6 centimetres tall to see the salt grain like this! Look at the little square in the middle of Fig. 6.3. The next time you see it, it will have been magnified a hundred times. The things in it will fill the whole of the next picture (Fig. 6.4).

Now we are looking at only a tiny part of the bottom corner of the salt grain. To see it like this you would need to be less than a fifth of a millimetre tall! Look at your ruler. Try to imagine being a fifth of a millimetre tall! At that height, the salt grain would look as big as a house.

One or two other interesting things have now become big enough to see:
a diptheria bacterium
a colon bacterium
a smallpox virus (the slightly larger fuzzy dot in the middle)
Remember, these things are $10 \times 10 \times 100$ times bigger than real life.

Notice that you still cannot see the salt particles – they are still too small. To be able to see them you would have to shrink another 1000 times! The salt grain would look about half the height of Mount Everest. We would have to blow up an area 1.5 mm square in the picture to make a new one 1.5 metres square. The total magnification would need to be 10 000 000 times, and the most powerful microscopes can only manage about 100 000.

6.2 Coloured lights

We cannot find out about atoms by looking at them – they are far too small. What can we do? There is a clue in Fig. 6.5.

Fig. 6.5

Rockets and roman candles are a familiar sight on the 5th of November. For thousands of years people have known how to colour their flames and explosions. Different chemical substances give different colours as they burn.

Many roads and motorways are now lit by sodium vapour lamps. These produce a bright yellow light. Electricity supplies the energy to the sodium vapour in the lamps.

Other gases can be made to give out light. Neon strip lighting uses this idea. Electric current passing through the gas produces a reddish glow.

Different substances produce different colours when they are heated. You can investigate it like this.

Investigation 6.1 Flame colours and atoms

You will need:
solutions of copper chloride, sodium chloride, calcium chloride, barium chloride, each in a small test-tube
a piece of nichrome wire joined to a glass rod
2 small test-tubes containing about 1 cm³ of dilute hydrochloric acid
test-tube stand
watch glass
spatula
Bunsen burner

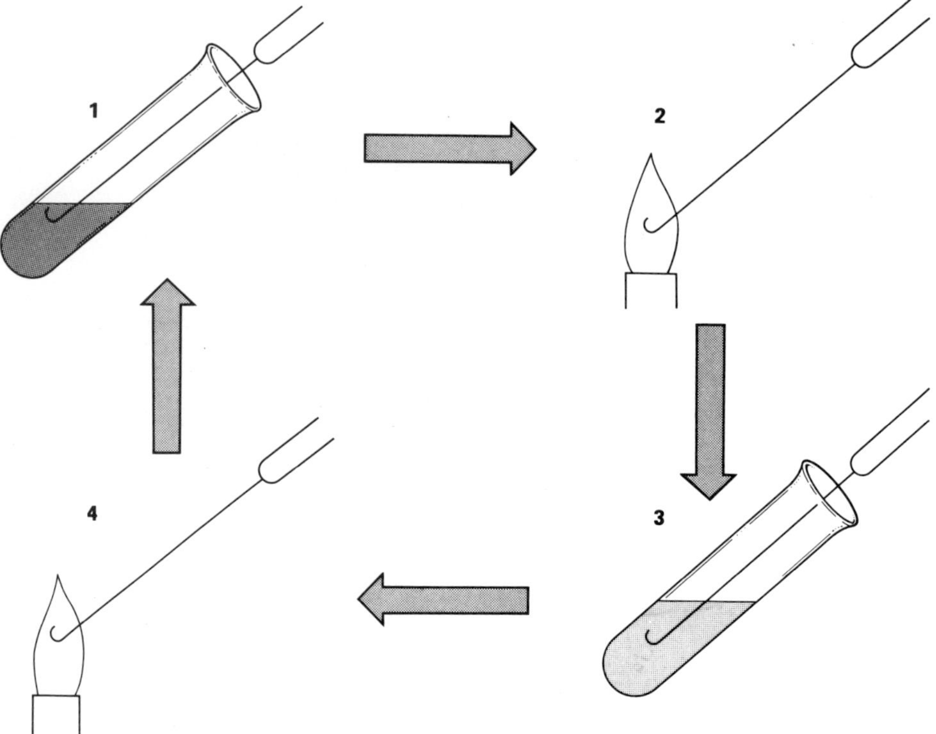

Fig. 6.6 Investigation 6.1

1 Dip the wire in the acid. Take care not to wet the glass. Heat the wire strongly in a non-luminous Bunsen flame.
2 Repeat this until the wire does not make any extra colour in the flame.
3 Dip the wire into one of the 'test solutions'.
4 Heat the wire in a non-luminous flame. Take care not to put the glass in the flame.
5 Clean the wire by repeating steps 1 and 2.
6 Repeat steps 3 and 4 with another test solution. . . . and so on.

Fig. 6.7

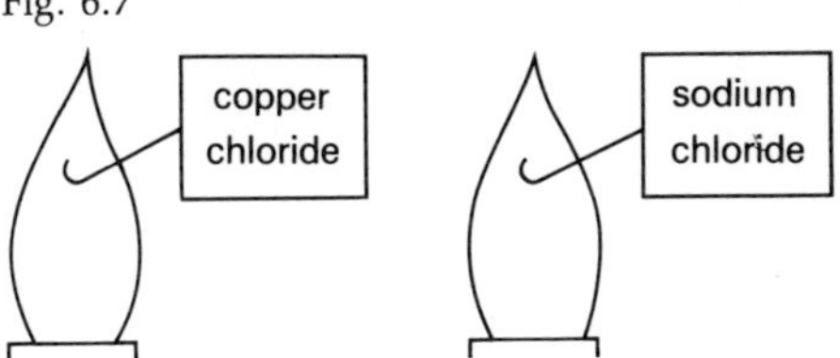

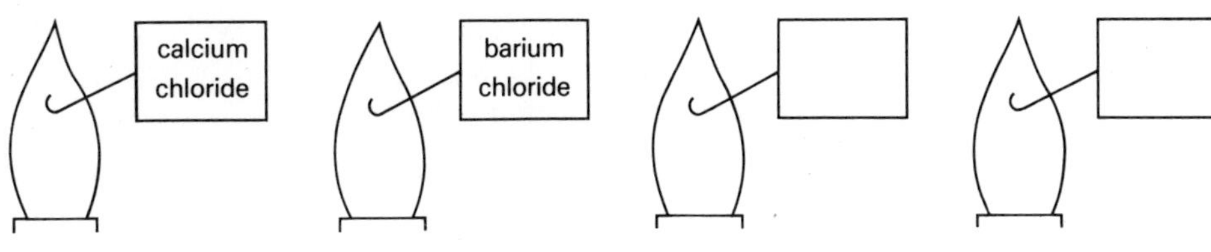

IN YOUR NOTEBOOK

✷ Write a heading: *Flame colours and atoms*

✷ Copy Fig. 6.7 and write the flame colours on the flames. Use coloured pens if you have them. Leave the last two diagrams blank. You will use them later.

Repeat Investigation 6.1 with these new solutions:
   copper sulphate
   copper nitrate

6.3 Patterns in the flame colours

What do the results of your investigations mean?

To answer that you need to think carefully about what you did and why. You took a very tiny bit of one of the test substances (probably less than a millionth of a gram) on the wire. You could think of this tiny bit as a mysterious 'Black box'. All you knew about it was its name. You could not see what it was like inside. (Neither could any other scientist!)

You heated it very strongly. When you heated it, it gave out coloured light. Each of the test substances gave light of a different colour. What was different about them? Of course, each of them had a different metal as part of it.

In the second part of the investigation, the test substances contained the same metal – copper. What flame colours did they produce?

IN YOUR NOTEBOOK

✷ Complete the last two diagrams you drew earlier.

✷ Write these statements, filling in the gaps:

Four test solutions contained a _______ of different metals. Each of these gave out l _______ of a different colour when placed in a flame.

Three solutions containing the atoms of the same _______ gave the same _______.

Atoms of different metals behave differently in a flame.

6.4 Elements

You remember (page 102) that the ancient Greeks thought up the idea of atoms. They imagined that all atoms were the same. Today's scientists have to believe that there are many different kinds of atoms. That is the only way they can explain things like the flame colours. You will come across more of their 'evidence' later.

How many different kinds of 'stuff' can you think of? How many different kinds of 'substance'? Have a go! Here is a start:

A aluminium
B butter
C chocolate
D dandruff
E Epsom salts

IN YOUR NOTEBOOK

✳ Write a heading: *Looking at different substances*

✳ Write your list of 'different substances'.

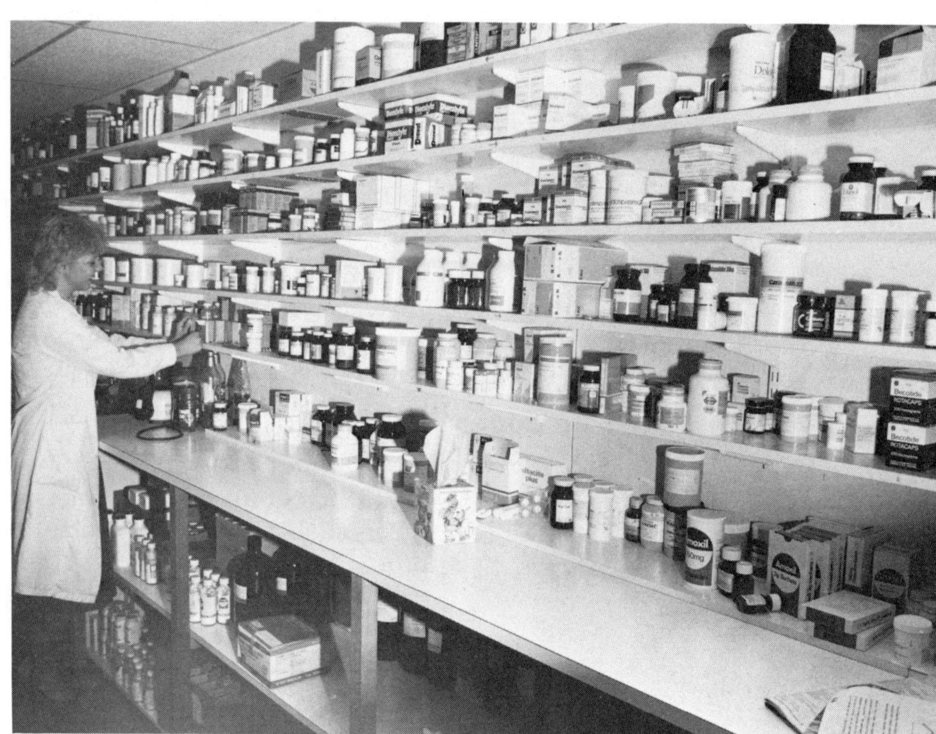

Fig. 6.8 Shelves of different substances in a chemist's dispensary

There must be millions of different substances in the world. All of them are made of atoms. How many different kinds of atom are there? Have a guess.

You will probably be surprised to find that they are all shown in Fig. 6.9. Count them. How many are there?

Substances which are made of only one kind of atom are called *elements*. So this table is a 'table of elements'. The different kinds of atom are shown by different code letters. Some of the code letters are easier to remember than others.

Fig. 6.9 The Periodic Table of the elements

6.5 Symbols

He stands for helium – the lighter than air gas used to fill balloons.
C stands for carbon – the 'lead' in your pencil.
O stands for oxygen – the gas that releases energy in our bodies.
But what about Na? What does that stand for?

Fig. 6.10 Airship filled with helium

Fig. 6.11 Sodium explodes and burns if it comes into contact with water!

6.6 A formula

Perhaps you know that the code Na stands for sodium. The element got its name from soda – a mineral used in washing and baking. The Latin name for soda was natrium.

Sodium is a metal – but not one you could make into knives and forks. It breaks up if it is left in the air, and it explodes and burns if it touches water.

Cl stands for chlorine. It is a greenish-yellow gas. It is used as a bleach and a disinfectant – in swimming pools for instance. It has been used in war as a poison gas.

The atoms of sodium and chlorine can be joined together to form a new substance. Can you guess what it is?

The answer is salt – the stuff you use to flavour food.

Fig. 6.12 Salt pans – small areas of sea water evaporating to produce salt

Scientists' name for salt is sodium chloride. The code for it is NaC1. Scientists call a code like this a *formula*.

IN YOUR NOTEBOOK

✳ Write these statements, filling in the gaps:

The world is made up of millions of different substances. The substances are made up of only _______ different kinds of atom. Substances made of only one kind of atom are called _______. There are _______ elements. Substances made up of atoms of at least two elements are called compounds.

Later on, you will learn about many more of the elements. And you will find out a lot more about formulae.

6.7 The combination game

How is it that all the millions of substances in the world can be made from only about 100 different kinds of atoms?

You will know the answer when you have played *the combination game*.

Rules

1 At least two players are needed.
2 Each player has three different coloured building bricks – the kind that click together (Fig. 6.13).
3 Each player has to put the bricks together to form 'combinations'.
4 A combination is any number of bricks (more than one) in any mixture of colours (Fig. 6.14).
5 Combinations can be different shapes (Fig. 6.15).

The winner is the person who makes and sketches (recording your results is as important as getting them!) the greatest number of combinations in five minutes.

Play the game a few times and compare your friends' combinations with yours. Have you made any that no one else thought of? How many different combinations do you think there are? Can you work it out?

Now you should be able to see how a hundred different kinds of atom can combine together to form the millions of different substances in the world.

Investigation 6.2 Making a real combination

You will need:
a little powdered sulphur
a little iron powder or powdered filings
3 pieces of heatproof paper (about 7 cm × 2 cm)

a pair of tongs
a Bunsen burner
a small magnet
a magnifying glass

1 Look carefully at each of your samples (sulphur and iron) in turn. Smell them. Test them with your magnet.
2 Put a little iron into a folded piece of heatproof paper. Heat it over the Bunsen flame. Watch carefully.
3 Repeat 2 with a little sulphur powder.
The fumes from burning sulphur can be dangerous. Be careful not to breathe any of them in.

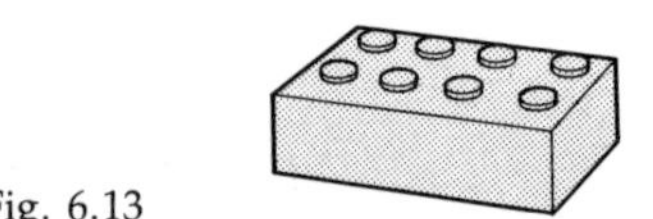

Fig. 6.13

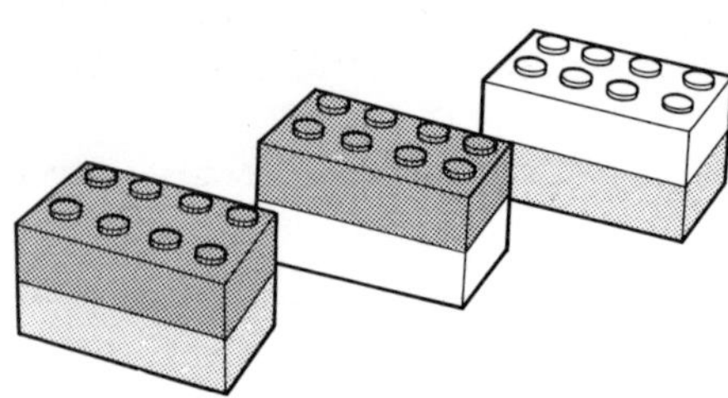

Fig. 6.14

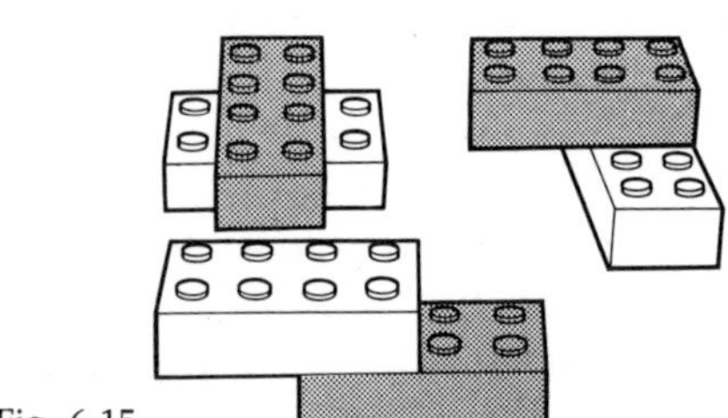

Fig. 6.15

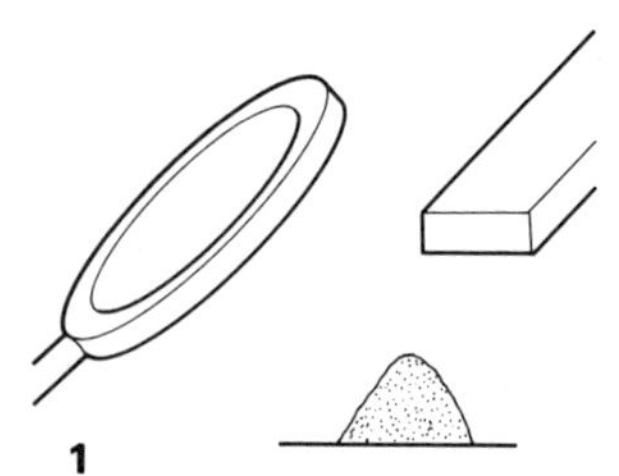

Fig. 6.16

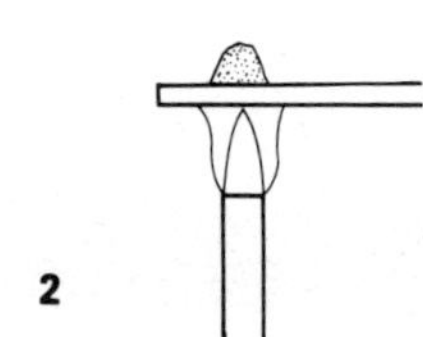

Fig. 6.17

IN YOUR NOTEBOOK

✻ Copy out this table and fill in the results of your observations:

Test	Iron	Sulphur
Appearance		
Smell		
Effect of magnet		
Effect of heat		

4 Mix equal amounts of sulphur and iron together. Heat the mixture. When things have started to happen, take the paper away from the heat. Watch carefully.
5 When the 'reaction' has finished, let the new substance in the paper cool off. Examine it as you did the powders, and put the results in your table.

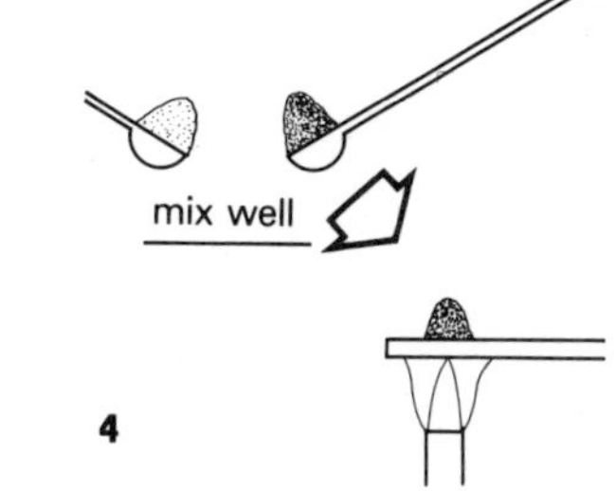

Fig. 6.18

IN YOUR NOTEBOOK

✻ Write these statements, filling in the gaps:

When a mixture of iron and sulphur is heated, it ________ . A compound is formed. It is called ________ ________. Its formula is ________.

Iron sulphide is found in nature, as iron pyrites. Fig. 6.19 shows what it looks like. Can you see why it is called 'fool's gold'?

Fig. 6.19 Iron pyrites

6.8 An order of violence

You have probably seen a set of experiments like the one shown in Fig. 6.20.

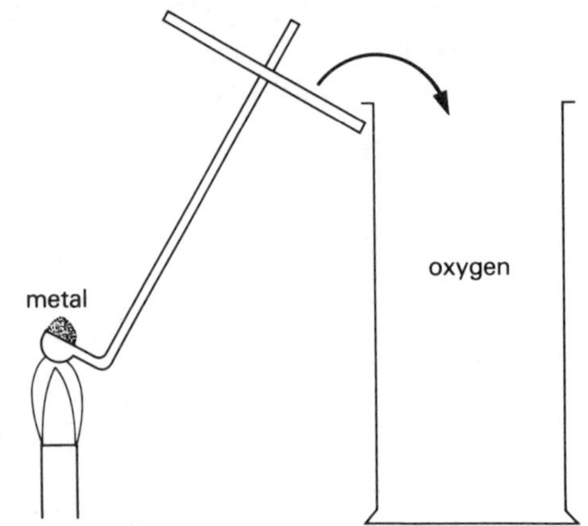

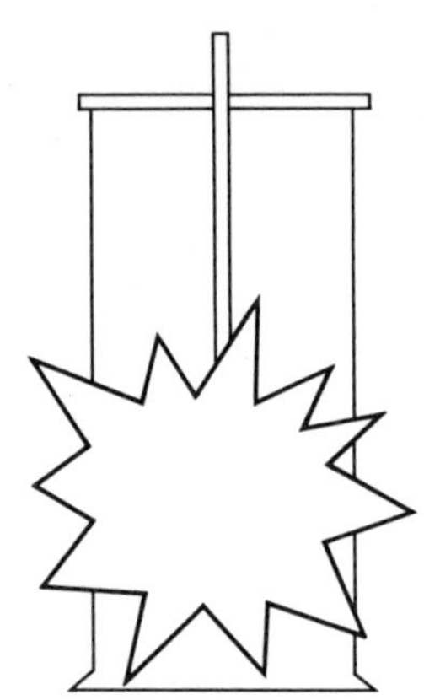

Fig. 6.20 Burning metals in oxygen

Different metals are burned in the gas oxygen. If you have seen it you will know that some of the metals burn furiously giving out a bright light, spluttering and sparking. Others hardly seem to burn at all.

As the metals burn, their atoms interact with the oxygen molecules and combine with them to form the compounds called *oxides*. *But some interact more strongly than others*. Aluminium is one of these. It interacts so strongly that it will tear the oxygen away from an oxide of another metal. A mixture of aluminium powder and iron oxide is used by British Rail to weld railway lines. When the mixture is set off with a fuse, the reaction is so violent that the ends of the rails are melted together. It is called the *Thermit* reaction. Your teacher might demonstrate it for you.

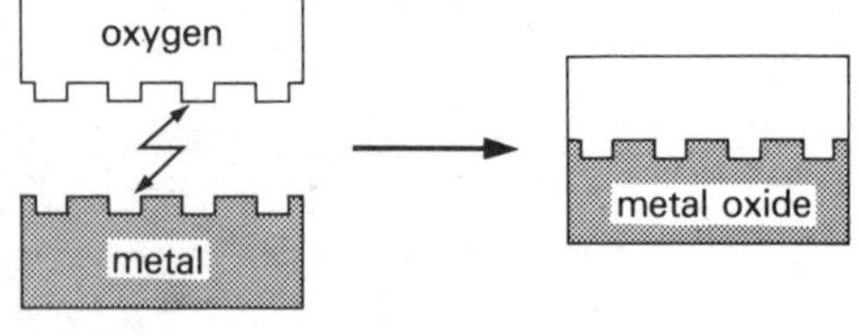

Fig. 6.21

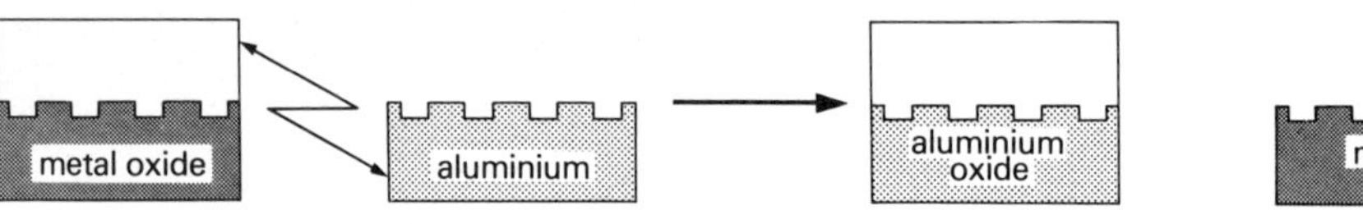

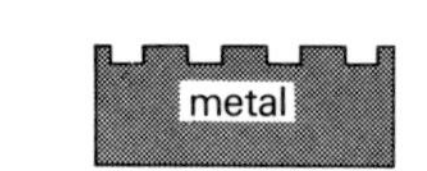

Fig. 6.22

The non-metal carbon interacts with oxygen more strongly than some metals and less strongly than others. For instance, it will tear the oxygen away from copper oxide to leave copper metal. This is the idea behind smelting (see page 41). But it will not do the same with aluminium oxide.

You will remember from page 40 that silver and gold can be found 'native' in small pure nuggets. Does this mean that these metals interact strongly or weakly with the oxygen in the air?

6.9 A closer look at sulphur

People have been using sulphur for thousands of years. Its common name is brimstone – from words meaning 'burn-stone'. People used to burn it 'to drive away evil spirits', to fumigate houses after illness, and to bleach fabrics. Nowadays it is one of the most important 'raw materials' of industry. Some of the ways it is used are shown in Figs 6.23–6.26. And, as you will see, it is very important to living things.

Fig. 6.23 The making of cellulose. Viscose is forced through a bath of sodium sulphate and sulphuric acid, and comes out like this.

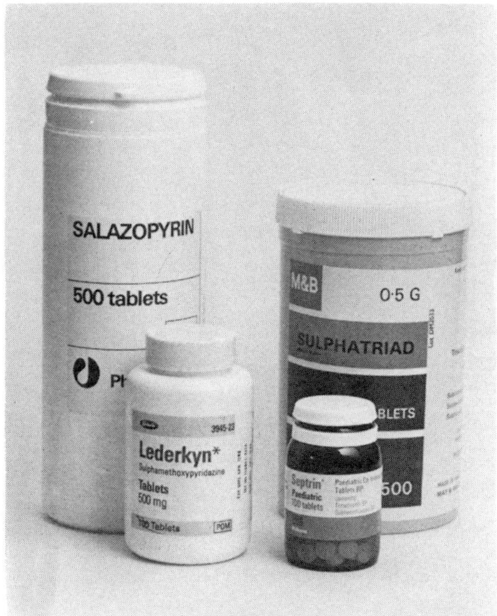

Fig. 6.24 Some medicines that have sulphur or sulphur compounds in them

Fig. 6.25 Some fertilisers contain sulphur

Fig. 6.26 Sulphur is used in making rubber for tyres

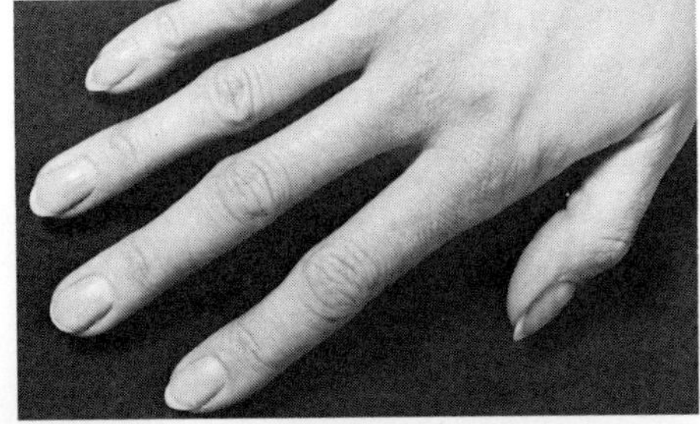

Fig. 6.27 Fingernails are made from keratin

Your skin cells die naturally all the time. (Most household dust is dead flakes of skin.) Some of the cells turn into keratin. Its molecules are like very long chains of different atoms. The chains are fastened to each other by little bridges made of sulphur atoms.

Keratin forms many useful parts of the body:

hooves, horns and wool   hair and nails
beaks, talons and feathers   shells
spines and claws   fur

Do you know someone who has had a 'perm'? Chemicals were poured on to

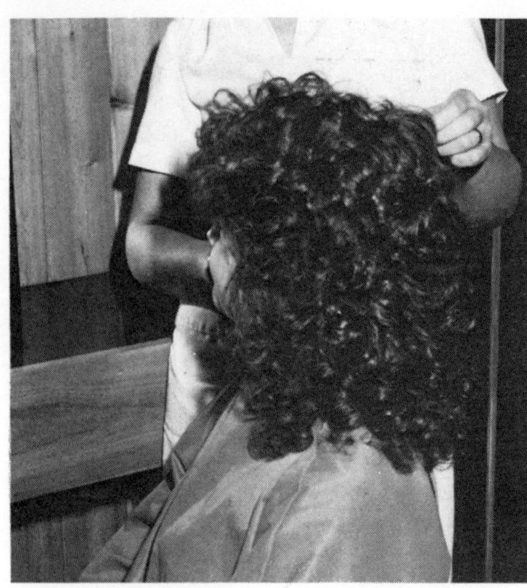

Fig. 6.28 This person has just had a 'perm'

the hair. They broke the sulphur bridges between the keratin chains. The hair became soft and easy to set. Other chemicals were used to put back the sulphur bridges to hold the keratin – and the hair – in its new shape.

Sulphur forms part of some of the most important body chemicals. These include substances which help you get energy from the food you eat, adjust the sugar balance in your blood, and make the proteins that build your muscles.

You would probably like to know where sulphur comes from. For part of the answer, read the story that follows.

6.10 The Herman Frasch story

On page 36 you saw that millions of years ago the remains of plants and animals were trapped under the ground when the sedimentary rocks were laid down. Microbes were trapped too and some of these lived by digesting the rock itself. One of the waste products of the microbes' activity was a gas called hydrogen sulphide – which smells of bad eggs. Other microbes helped to convert this gas into the mineral element *sulphur*.

Fig. 6.29 Some microbes live by digesting rock!

Millions of years after the microbes had done their work, a seventeen-year-old called Herman Frasch arrived in the USA from Germany. His aim was to get rich as quickly as possible. He was very bright and worked very hard. Within a few years he had set up his own laboratory in Philadelphia and had become an expert oil scientist.

By the time Herman had set up his lab, there were many oil wells in the area. Oil had been struck about twenty years before. A few people had sold it as medicine (!) and some was used as a lubricant but it was unpopular for burning – mainly because of its awful smell. Herman worked out that the smell was caused by sulphur in the oil and found a way of separating it out. He patented his discovery and began to make a lot of money from making the oil smell nice! He found that he could sell the sulphur too.

Herman began to think of other ways of getting hold of sulphur. He knew there was plenty of it a few hundred miles away under the swamplands of Louisiana, but it was deep and difficult to reach. He thought: 'If only I could pump the sulphur out like oil.' People made fun of the idea – sulphur is, after all, a solid mineral. But Herman was not discouraged. His aim was to find a way to melt the sulphur underground and then pump it up. How would you have done it?

Herman hit on the idea of pumping very hot water down a hole to melt

Fig. 6.30 An oil well of Frasch's time

the sulphur and float it up. He had dozens of problems to solve before he could make his idea work. Quicksand and water filled in the bore holes, and pockets of poisonous gas came bubbling up. Worst of all, he could not get enough fuel to heat his water.

Then, just as he made his method work for the first time, oil was struck in nearby Texas. Suddenly he had more fuel than he needed. His fortune was made!

6.11 Herman Frasch's famous sulphur well

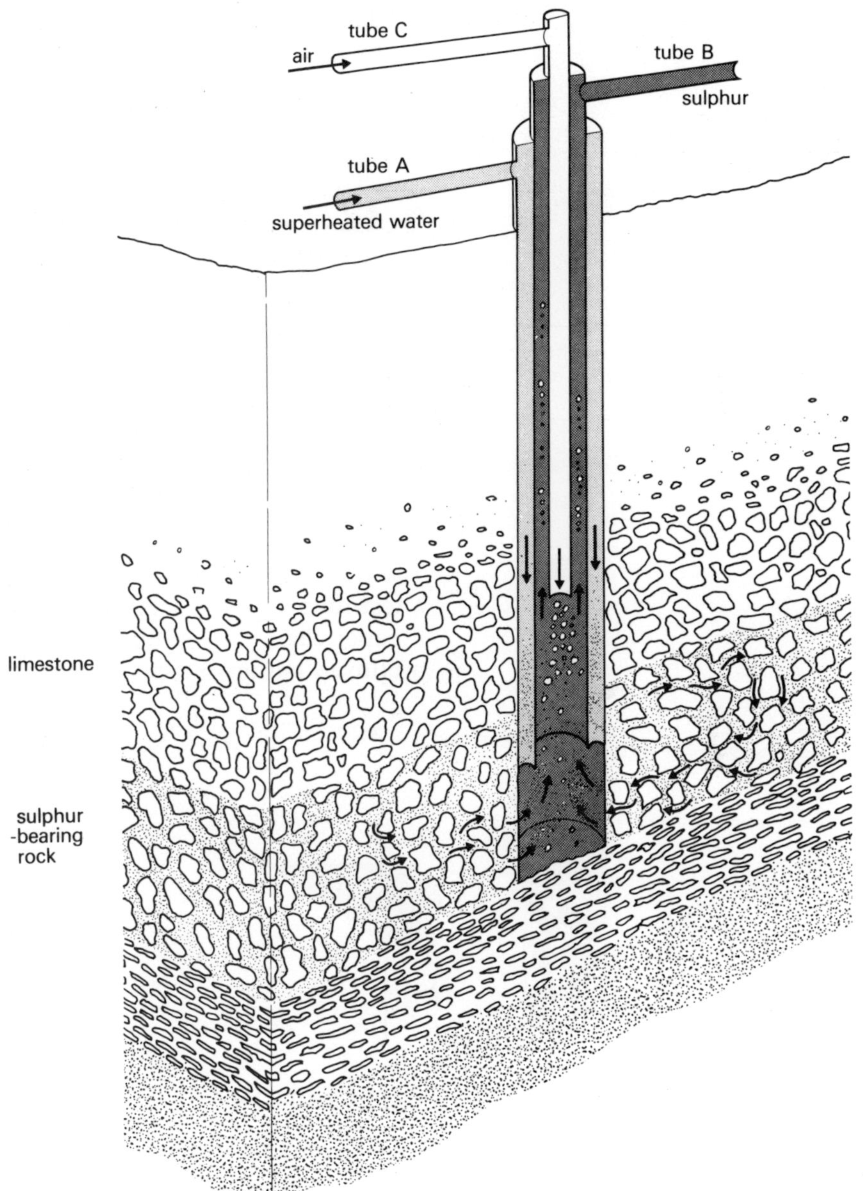

Fig. 6.31 Herman Frasch's sulphur well

Fig. 6.31 shows a cut-away view of Herman's clever invention. Its secret is three tubes nesting inside each other. It works like this:

1 Very hot water ('superheated' under pressure) is pumped down tube A.
2 The sulphur melts and starts to come back up tube B.
3 Compressed air is blown down tube C, the sulphur froths and floats up tube B.

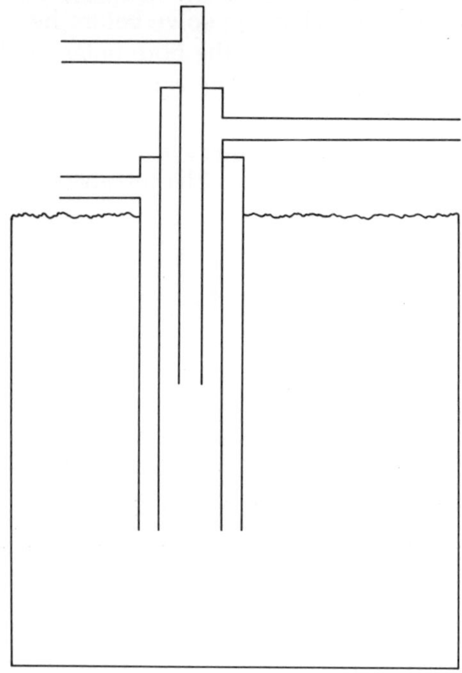

Fig. 6.32 Which ways do water, sulphur and compressed air move in this diagram of a Frasch well?

Fig. 6.33 Investigation 6.3

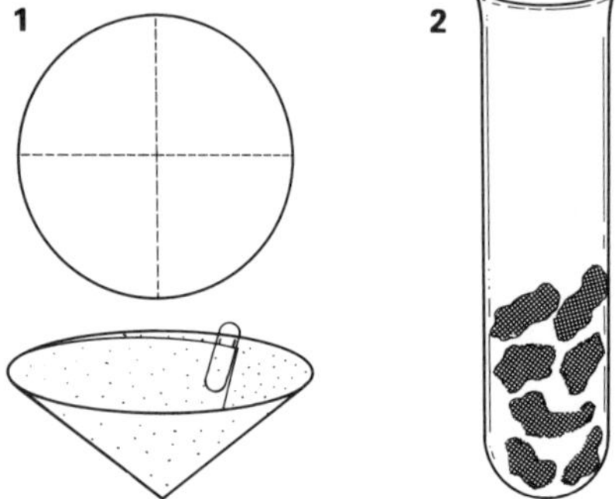

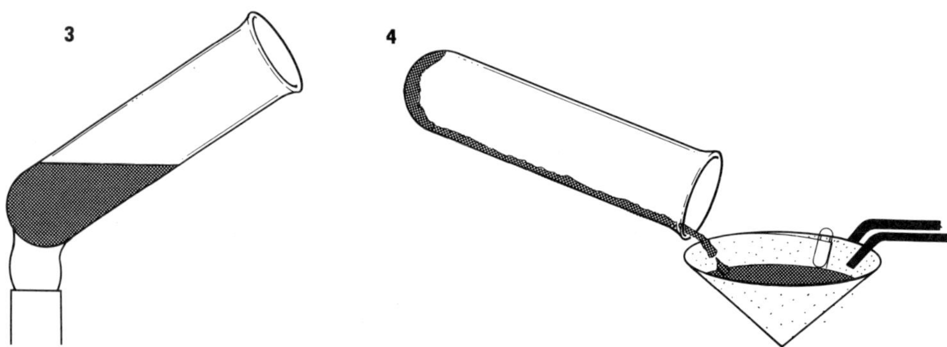

IN YOUR NOTEBOOK

* Write a heading: *Looking at sulphur*

* Draw some pictures to show how sulphur is used in industry.

* Make a list of *'Uses of sulphur'*.

* Write a few words starting with:
  Sulphur is important to living things. It forms part of _______.

* Copy the diagram of a Frasch well in Fig. 6.32. Put arrows on the diagram to show the movement of water, compressed air, and sulphur in the three stages.

6.12 Different kinds of sulphur

Sulphur is an element. That means, of course, that it is made of identical sulphur atoms. So it comes as a surprise to most people to find that there are different kinds of sulphur. The instructions below show how to make samples of two different kinds.

Take care not to breathe the fumes from the hot sulphur in any of the preparations and take care not to get hot sulphur on your skin – it can give you a nasty burn.

Investigation 6.3 Making different kinds of sulphur

You will need:

a test-tube   small pieces of sulphur
a paper clip   a filter paper
a Bunsen burner   a beaker of cold water

PREPARATION 1

1 Fold a filter paper into a cone and fasten with a paper clip.
2 Half fill a test-tube with small pieces of sulphur.
3 Warm gently in a small flame until the sulphur has just melted and is runny. Move the tube about so that the sulphur is heated evenly.
4 Hold the paper cone with tongs. Pour the liquid sulphur into the paper cone.
5 When a crust forms on the sulphur, remove the paper clip and open the cone.
6 Open the paper out and look carefully at the crystals which have formed in the cone. *Careful! There will still be hot, sticky sulphur in the cone.* Compare the crystals with the shapes in the diagram. Which shape can you see?

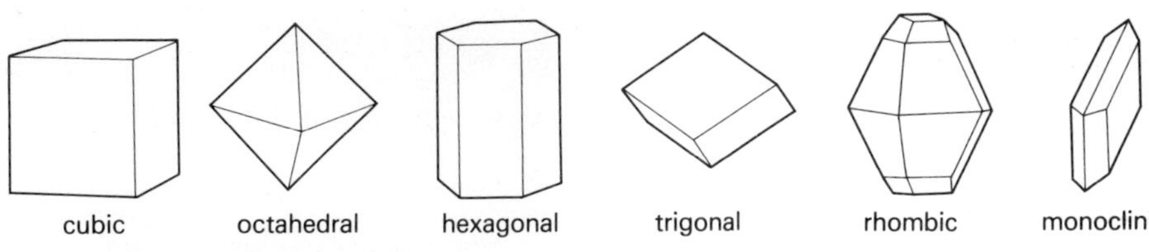

Fig. 6.34 Different examples of crystals

PREPARATION 2

Repeat Preparation 1, steps 2 and 3. Keep tilting the tube a little to see how the sulphur runs. Watch carefully what happens. Make a note of anything odd.

When the sulphur boils, pour it into a beaker of cold water. Pour the water off straight away, and take the sulphur in your fingers. *Careful! It might still be hot.* Squeeze the sulphur and pull at it with your fingers.

IN YOUR NOTEBOOK

✳ Write a heading: *Making different kinds of sulphur*

✳ Draw diagrams to show the two preparations.

✳ Sketch any crystals that you saw.

In Preparation 1 you made sulphur in the form of crystals. Preparation 2 gave you a kind of sulphur that is stretchy and rubbery. As you were heating the sulphur you probably noticed that the liquid suddenly became thick. Then it got thinner until it boiled.

How do scientists explain these changes? Remember that there is only one kind of sulphur atom.

The answer has to be that the different kinds have different arrangements of atoms. In fact, scientists have found evidence that the building block in sulphur crystals is a ring of eight sulphur atoms, called an S_8 ring. The ring is not flat but buckled. It looks like Fig. 6.35.

It is easier to see how the atoms are arranged if you imagine them to be shrunk into balls and sticks as in Fig. 6.36.

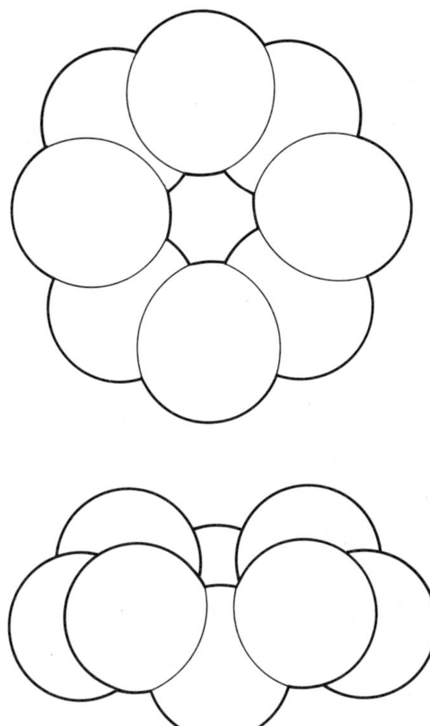

Fig. 6.35 A ring of eight sulphur atoms (an S_8 ring)

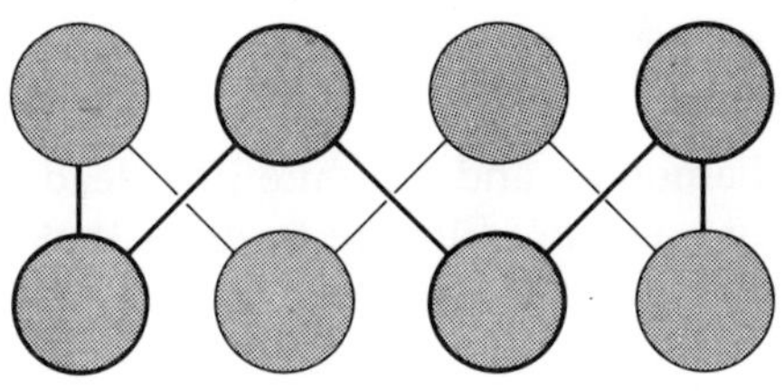

Fig. 6.36 An S_8 ring drawn as 'balls and sticks'

6.13 What happens when you heat sulphur?

Scientists are far from certain about the answer to this question. One explanation goes as follows.

The Bunsen flame gives energy to the S_8 rings. They vibrate faster. Then the rings separate and the solid turns to liquid. The rings can move about freely so the liquid is easy to pour.

More heating causes the liquid to thicken. This is because the S_8 rings break open and form chains. It happens at a definite temperature (160 °C). The chains, which can be a million atoms long, get tangled with each other.

More heating breaks up the chains, some S_8 rings form again, the liquid becomes thinner. Finally some of the S_8 chains split up into S_2 units and the sulphur becomes a vapour.

The stretchy and rubbery sulphur from your beaker of water is called *plastic* sulphur. It is a mixture of the giant chains and some smaller units (including a few S_8 rings). The smaller units help the chains to slide over and round each other.

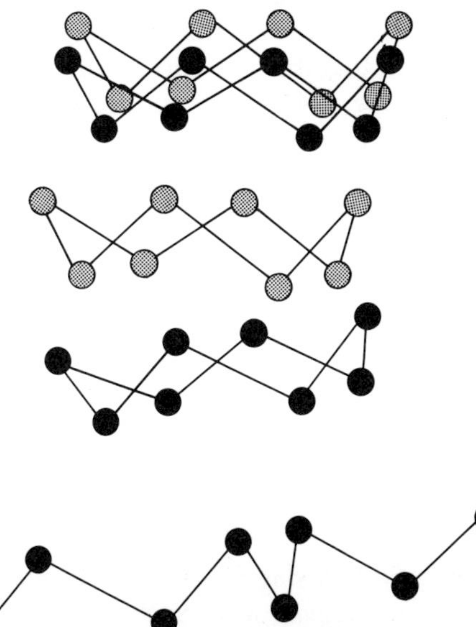

Fig. 6.37 Heating sulphur. The black and grey circles are sulphur atoms in different rings

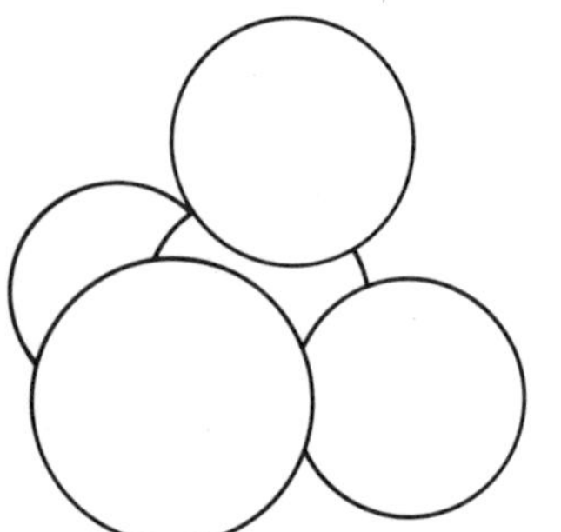

Fig. 6.39 The arrangement of atoms in diamond

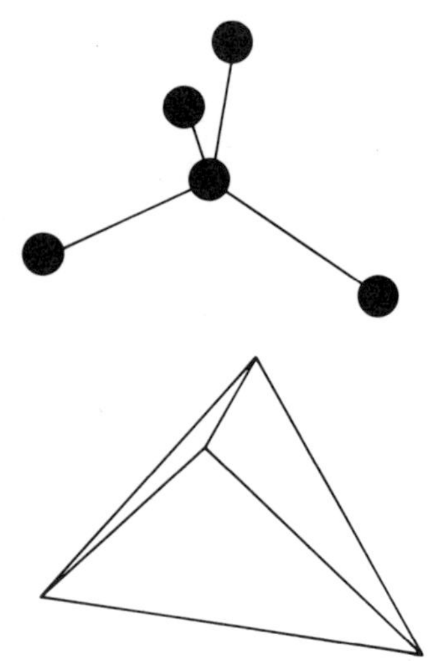

Fig. 6.40 The atoms in diamond shrunk into 'balls and sticks'

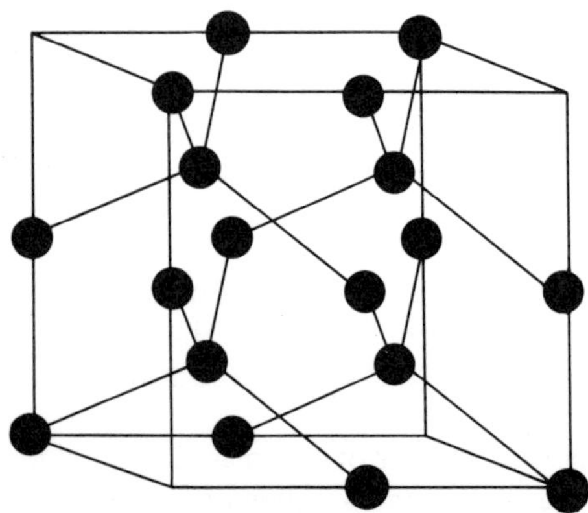

Fig. 6.41 The zig-zag structure of a diamond crystal

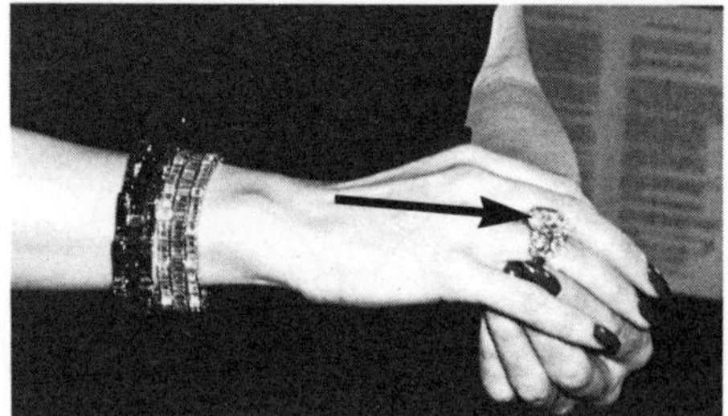

Fig. 6.38

Look at the different kinds of stuff marked with arrows in Fig. 6.38. What have they in common? Can you say?

6.14 An atomic look at carbon

You probably know that charcoal, soot, diamonds, and graphite (the 'lead' in a pencil) are all forms of the same element – carbon. Strange to think that a diamond and a piece of pencil 'lead' are made out of the same kind of atoms isn't it! The reason for their differences is the same as for the different kinds of sulphur. Their atoms are arranged in different ways. They have different 'structures'.

Scientists believe that the atoms in diamond are arranged as in Fig. 6.39. (Imagine millions of millions all fitted together like the few in the diagram.) To get a clearer idea of the way they are fitted together we can 'shrink' the carbon atoms into 'balls and sticks' (Fig. 6.40). Fig. 6.41 shows the zig-zag structure of a diamond crystal. It is equally hard in all directions.

IN YOUR NOTEBOOK

✳ Write a heading: *Arrangements of carbon atoms*

✳ Copy or trace Figs 6.40 and 6.41.

✳ Write these statements, filling in the gaps:

In the diamond structure, every carbon atom touches _______ others. Any five carbon atoms make a _______ shape. The shape has _______ faces. The shape of each face is a _______.

✳ Write out this important piece of information: *A single diamond crystal is made up of millions of atoms joined together in a regular way. Such an arrangement of atoms is called a 'giant structure' (sometimes a 'macro-molecular' structure).*

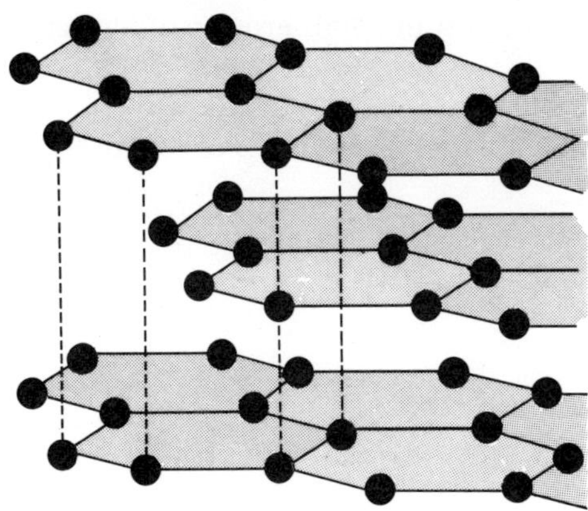

Fig. 6.42 Graphite – how does its structure differ from that of diamond?

Fig. 6.42 is a diagram of graphite. The atoms have been shrunk into 'balls and sticks'. Can you see how its structure differs from that of diamond?

Can you see that the atoms are arranged in layers – rather like a pack of cards? Can you guess why graphite has a slippery feel? What is happening to the layers of atoms when you rub graphite between your fingers?

IN YOUR NOTEBOOK

✳ Write these statements, filling in the gaps:

In the graphite structure, each carbon atom touches ______ others. The carbon atoms make ______ shaped patterns. They are arranged in ______. The layers are loosely connected and so can slide over each other. This explains why graphite feels ______.

✳ Finally, write out this important pattern:

A number of elements, for example carbon, and sulphur, exist in several different forms. These forms are called allotropes.
The different properties of allotropes are explained by different arrangements of their atoms.

6.15 The MP–BP pattern

What does it mean to say that a substance is *pure*? Can you explain what the words 'a pure substance' mean in terms of molecules and atoms? Have a go.

Scientists can tell whether a substance is pure or not by measuring the temperature at which it melts or boils. Pure substances have exact melting and boiling temperatures (called melting points, MP, and boiling points, BP).

There are books full of tables of MP's and BP's for almost every known substance. Here is part of one of the tables.

Element	MP(°C)	BP(°C)
Aluminium	659	2447
Copper	1083	2582
Iodine	114	183
Iron	1539	2887
Phosphorus	44	281
Silicon	1410	2667
Sulphur	119	445

IN YOUR NOTEBOOK

✳ Copy out the temperature chart in Fig. 6.43.

✳ Write the names of the elements next to their MP's and BP's. Roughly in the right place will do. Copper has already been filled in for you.

When you have filled in the chart, you will need two more pieces of information to see a pattern. Here they are:

Aluminium, copper, iron, and silicon are made up of separate atoms linked together in a giant structure.

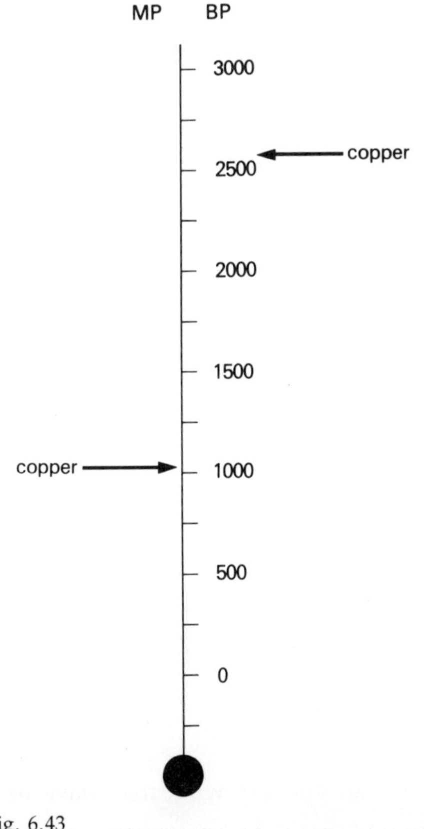

Fig. 6.43

Iodine, phosphorus, and sulphur are made up of small groups of atoms – that is, they are made up of molecules.

Can you see the pattern?

The pattern works for compounds as well as for elements. For example:

Compound	MP(°C)	BP(°C)
Sodium chloride	808	1465
Carbon dioxide	−56	−78

Which compound has a giant structure?

IN YOUR NOTEBOOK

✻ Write a heading: *A pattern of melting points and boiling points*

✻ Write these statements, filling in the gaps:

Substances which are giant structures have _______ melting points and boiling points.
Substances which are molecular structures have _______ melting points and boiling points.

You can explain the MP/BP pattern by thinking what might happen to the molecules when a material melts or boils.

On page 92 you saw that the molecules of a solid have very little freedom to move – they are 'locked' in position amongst their neighbours. Those of a liquid have more freedom but are still partly 'locked up'. On page 91 you saw that the higher the temperature, the faster the molecules move.

If you heat a solid, the molecules will vibrate faster and faster until they reach the point where they partly break free from each other. That is when the solid melts, becoming a liquid. Heat the liquid and you will reach a point where the molecules are moving fast enough to burst out of the liquid and go off on their own. Then the liquid will be boiling – changing into a gas.

It is not difficult to see why giant structures have higher MP's and BP's than molecular structures. The atoms of a giant structure – all tightly locked together – need to be moving much faster before they can break free.

6.16 Can you use the pattern?

So-called tar sands cover large areas in Canada. Tar sands are made up of grains of sand coated with a very thick, sticky oil. Recently, scientists have been looking for ways to extract the oil. You can guess why!

The sand is mainly made up of a substance called *silica*. It is a giant structure made up of silicon and oxygen atoms. The oil or tar is a mixture of molecular substances.

How do you think the mixture might be separated? Can you use the MP/BP pattern to give you the answer?

PICTURE QUIZ

Look carefully at the pictures in Fig. 6.45. Can you say what they have in common?

Fig. 6.44 Tar sands in Canada

(a)

(b)

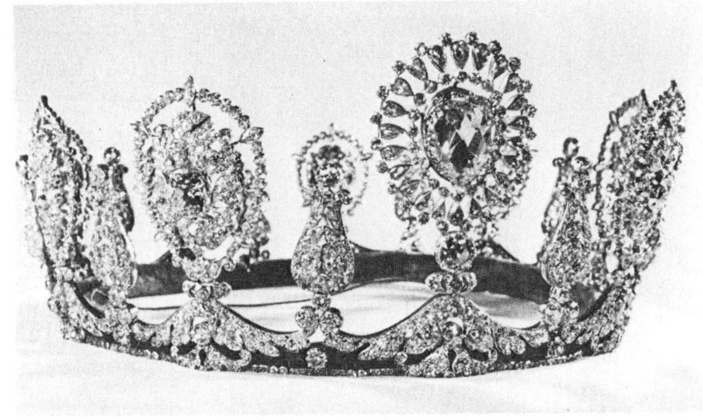

(c)

(d)

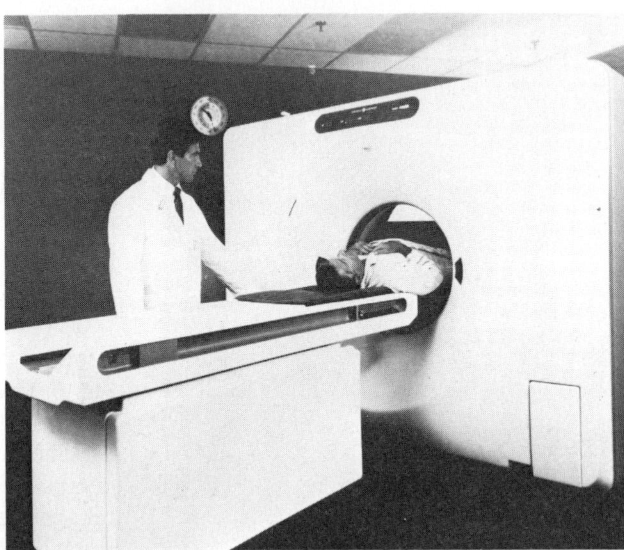

(e)

Fig. 6.45 (a) Stylus for a record player
          (b) Quartz watch
          (c) Jewelled crown
          (d) A 'squeeze-powered' gas lighter
          (e) A body scanner

6.17 Crystals

The pictures on page 119 show inventions which have one thing in common. They are all built around, and make use of, crystals. Did you guess?

Some crystals can change X-rays into light. Thousands of them are used to make the 'detector plates' of a body scanner. The pictures the scanner takes look like slices through the patient's body.

The table below lists some other properties of crystals.

IN YOUR NOTEBOOK

✱ Copy out the table and fill in the empty spaces in the right hand column. In each space write the name of something that uses the property. Some items are shown in Fig. 6.45. The first space has been filled in to give you a start.

Crystal property	Item which uses property
Flat sides, sharp edges, shininess, sparkle	Jewels
Very sharp corners, often very hard	
Can vibrate and give out very regular bursts of electricity	
Can change the size of electric currents	
Can change pressure into sparks of electricity	
Can change X-rays into light	
Can change a flash of light into an ultra powerful beam	

You can grow your own metal crystals in the way illustrated in Fig. 6.46. Leave the tubes in a rack where they will not be disturbed and look at them from time to time for the next day or two.

6.18 What is a crystal?

Scientists believe that a crystal is a collection of particles such as atoms arranged in a very regular way.

Of course, you cannot see atoms even through the most powerful microscope – they are far too small. But microscope pictures like the one in Fig. 6.47 suggest that the atoms must be packed together in a regular way. Can you see how it suggests that?

Scientists use 'bubble rafts' to work out theories about the ways that crystals form. They study breaks in the regularity of the bubble patterns. The breaks are called *dislocations*. Can you see them in Fig. 6.48? Pick a small area of the bubbles and look for wavy lines in the pattern.

You can make your own bubble raft with the apparatus shown in

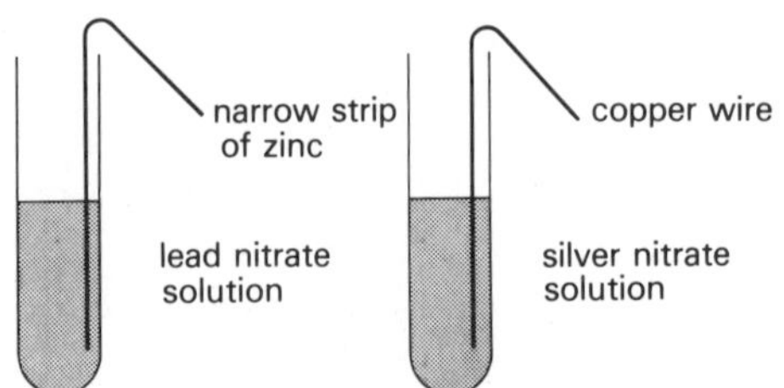

Fig. 6.46 Growing metal crystals

Fig. 6.47 Scanning electron micrograph of lead tin telluride crystals, magnified 30 times

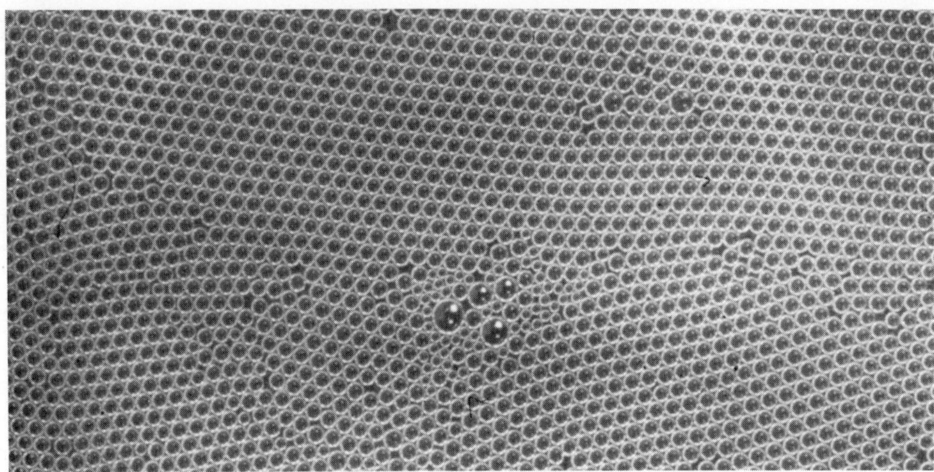

Fig. 6.48 A bubble raft – can you see breaks in the regularity of the bubble patterns?

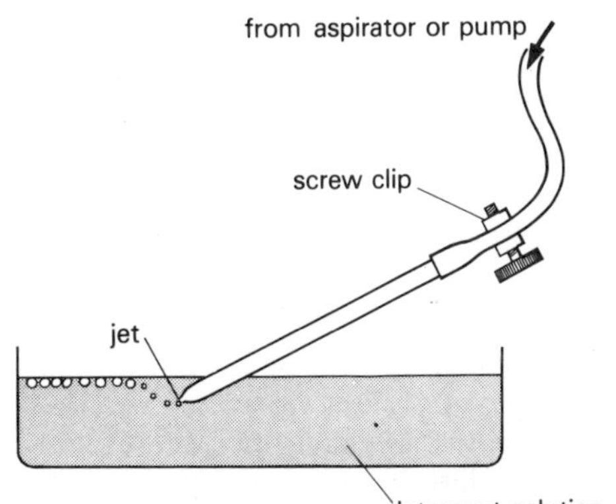

Fig. 6.49 Home-made bubble raft apparatus

Fig. 6.49. Adjust the flow of gas so that very small bubbles are formed very slowly. Try to cover the surface with a single layer of bubbles.

Can you see any parts of the raft where the bubbles are arranged in a regular pattern? Each bubble represents an atom of metal. Regular arrangements of bubbles represent regular arrangements of metal atoms – 'crystalline' parts of the metal. Your bubble-raft is a two-dimensional arrangement. You need to remind yourself that the atoms in a metal are in a three-dimensional arrangement.

6.19 Crystals of 'compounds'

So far, you have looked at crystals of sulphur, carbon (diamond and graphite) and some metals. In each of these crystals, only one kind of building block was used. The crystals were built up from the atoms of each element.

Fig. 6.50 shows some pictures of crystals made up of more than one kind of building block.

Fig. 6.50 (a) Calcite (calcium carbonate) crystals   (b) Alum crystals

Crystals of calcite are made up of atoms of calcium, carbon and oxygen, some combined into molecules.

Crystals of alum are made up of atoms of potassium, aluminium, sulphur and oxygen, some combined into molecules.

You can make some of these 'compound' crystals.

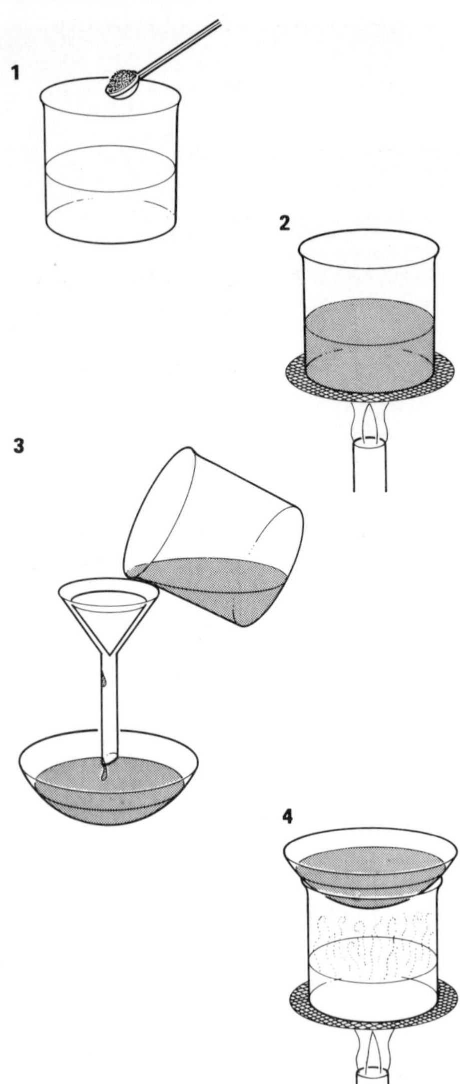

Fig. 6.51 Investigation 6.4

Investigation 6.4 Making compound crystals

You will need:
the reagents in the table below
a filter funnel
filter paper
a beaker

an evaporating basin
a spatula
a Bunsen burner

To make:		Materials ('Reagents')
Copper II sulphate	you need	sulphuric acid + copper carbonate
	or	sulphuric acid + copper II oxide
Zinc sulphate	you need	sulphuric acid + zinc
	or	sulphuric acid + zinc oxide
Magnesium sulphate	you need	sulphuric acid + magnesium
	or	sulphuric acid + magnesium oxide
Sodium chloride	you need	hydrochloric acid + sodium carbonate

1 Add 2 spatulas of the solid reagent to about 25 cm³ of dilute acid in a small beaker.
2 If there is no sign of any change, gently warm the mixture for a few moments. Set the beaker aside to let the reaction go on.
3 Filter the mixture. Collect the filtrate in an evaporating dish.
4 Warm the evaporating dish and its contents over a beaker of boiling water until there is only about half as much liquid left. Turn out the Bunsen burner.
5 Leave the dish to stand overnight. Carefully pour away the liquid from the crystals.
6 Gently press the crystals between sheets of filter paper to dry them. Use a hand lens to look at them.

Are your crystals like any of the shapes in Fig. 6.34, page 115? Are they coloured? Can you draw them?

Checkout

Keywords

allotrope
atom
boiling point
compound
crystal
element
formula
giant structure

kinetic theory
melting point
metal
mixture
molecule
smelting
structure
substance

Patterns

1 Special substances called elements can be made to react together to produce new substances called compounds.
2 When heated in a flame, compounds of different metals give out light of

different colours. A particular colour can be matched with a particular metal.

3 Compounds behave differently from the elements which make them up. (They have different properties.)

4 Some metals interact more strongly with oxygen than others. They can be arranged in a 'reactivity series'.

5 Some elements (sulphur and carbon for instance) can exist in different forms which behave in different ways. (Allotropes)

6 Pure substances have exact melting points and boiling points. Substances with 'giant structures' have higher melting points and boiling points than substances with 'molecular structures'.

7 Crystals have regular shapes, flat sides, and sharp edges. These things give clues to their structure.

A theory

Many scientific observations can be explained by believing that all substances in the universe are made up of extremely tiny particles, which we call atoms. The theory says that:

(a) There are about 100 different kinds of atoms.

(b) Some substances are made up of only one kind of atom (elements); others are made up of different atoms 'joined together' (compounds).

(c) Substances are made up of *either* single atoms all joined together to form 'giant structures'; *or* small groups of atoms called 'molecules'.

(d) Some elements can exist in the different forms called 'allotropes' because their atoms can be arranged in different ways.

True or false?

1 Scientists know about atoms because they can see them through the most powerful microscopes.

2 Every metal produces a different colour in a flame test.

3 The formula for a compound shows the way its atoms are arranged.

4 All the following are metals: calcium, carbon, chromium, and copper.

5 All the following are non-metals: chlorine, iodine, phosphorus, and sulphur.

6 Sodium metal has a giant structure but sodium chloride is molecular.

7 If you grind up some iron sulphide and stir it with a magnet you can get most of the iron out of it.

8 The carbon atoms in diamond are a different shape from those in graphite.

9 Statement B (below) is a correct explanation of statement A:
A. If you heat a salt solution, the water evaporates but the salt does not – so the salt is left behind.
B. This is because water has a giant structure and salt is molecular.

10 A laser makes use of the fact that some crystals produce electricity when light shines on them.

Problems

1 *Read this passage and try to answer the questions.*
One day, a scientist went back to the lab to find that there had been a flood. The labels had been washed off five bottles containing white powders. The scientist knew that the bottles contained various metal compounds. The problem was: which metal was in which bottle?

(a) What simple test would you suggest to solve his problem?

(b) How could he tell which bottle contained salt? (Tasting is *not* allowed.)

2 *Read the passage and try to answer the questions.*
In 1851, a church organ was being repaired in the town of Zeitz in
Germany. The pipes of the old organ were made of almost pure tin. The
workmen found that one pipe was covered with grey swellings like warts. If
rubbed, these 'warts' crumbled into a grey dust. Scientists of the time were
puzzled. Their only suggestions was that the 'tin pox' might be caused by
vibration of the pipes when they sounded.

Fifteen years later, after the very cold winter, a Russian scientist was
called in to investigate some stocks of tin in a warehouse. Some of the tin
bars had broken down into a grey powder. The scientist took samples of the
powder and heated it. It melted and, on cooling, was 'normal' tin again.

(a) What do you think the warts and grey powder were made from?
Explain how you get your answer.
(b) Thinking about *atoms*, how was the grey powder different from the
tin bars?
(c) If you had been the Russian scientist, what other experiments would you
have tried?

7 Energy

7.1 Human beings – the great tool-makers

We are not too sure when the first humans appeared on the Earth. It was probably sometime between three and four million years ago. Half a million years ago our ancestors probably looked like the people in Fig. 7.1.

Fig. 7.1 Our ancestors of half a million years ago may have looked like this

When you think about these early people, it seems surprising that they survived, and even more surprising that they became the most powerful animals on the Earth. They had less hair and were less resistant to cold than most mammals. They could neither run as fast nor kill as quickly as the cat family, they were less agile than the monkeys, and much weaker than the bears. So how did they survive? Why did they do so well?

They had three big advantages. First, they could touch the tip of any of their fingers with the tip of their thumb – something a chimpanzee cannot do. It seems a small thing, but it meant they could hold things better than other animals. It gave them the 'precision grip' shown in Fig. 7.2.

Second, they had learned to walk upright – which freed their hands to hold stones, sticks, clubs, and then spears. Third, they learned to make many different kinds of hand-held tools for fighting and hunting, for making clothes and shelters and, much later, to help them grow food.

Special amongst all the animals, humans became the great tool-makers.

Nowadays we live in a world full of complicated tools with many moving parts. We call them 'machines' and 'engines'. We all depend on them, and

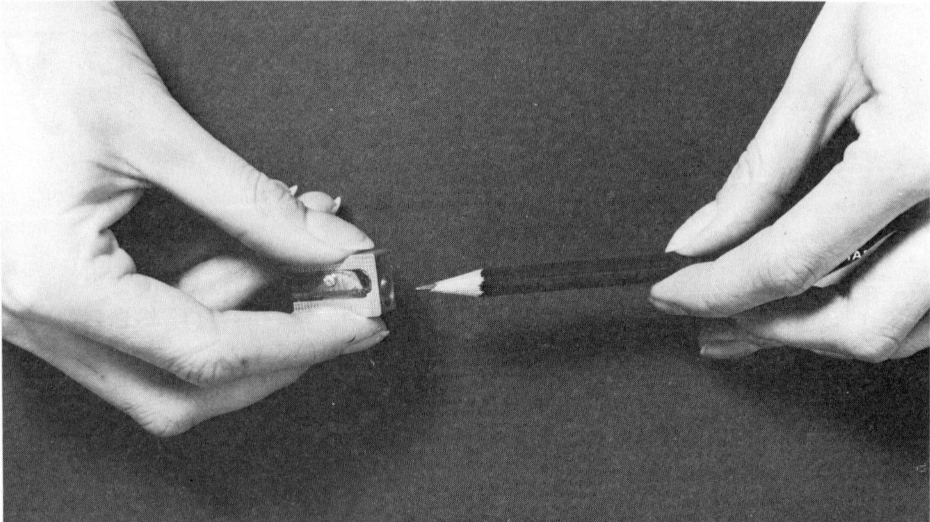

Fig. 7.2 The precision grip of humans – chimpanzees cannot do this

Fig. 7.3 We all depend on complicated 'machines' and 'engines'

many people spend their lives designing them, making them, or keeping them working.

This chapter sets out to answer questions such as: What is a machine? What is an engine? What do machines and engines do? What do they all have in common? Are there patterns in the ways that all machines and engines work?

On the way, the chapter looks at one of the most important ideas in science: the idea of *energy*.

One answer to the question: What do machines and engines *do*?, is that they make life easier and more convenient. It is easier and handier to use a machine than to do a job with your bare hands.

IN YOUR NOTEBOOK

* Write a heading: *Machines and engines*

* Copy the table opposite – but do not rule off the bottom. Copy the list of machines and engines into the left hand column. Then try to fill in the other two columns.

Machine/engine	How it makes life easier or more convenient	My estimate of the number of people needed to do the same job in the same time
Car jack	Needs much less force to lift car.	At least 4
Can opener (opening 5 cans)		
Egg whisk (beating up 6 eggs)		
Typewriter (typing 40 words)		
Lawn mower (hand operated)		
Washing machine		
Vending machine		
Chain saw		
Car wash		
Mechanical shovel (JCB)		

* When you have finished, think of one or two machines or engines of your own. Add them to the table.

* Try to find one or two pictures of machines and engines in newspapers and magazines. Stick them into your book.

* Choose one of the machines in your table and design an experiment to test your estimate in the right hand column. Talk about it with your teacher. See if you can try it out.

To see the patterns in the ways that machines and engines work you need to understand some important scientific ideas – the ideas of *inertia*, *force*, and *energy*. Start by looking at inertia.

7.2 Naturally lazy!

How would things move if there was no friction? What is the 'natural' way for things to move?

People argued about these questions for hundreds of years until, about three centuries ago, the great English scientist Isaac Newton (later, Sir Isaac) came up with an answer. His answer made clear one of the most important ideas in science – the idea of *inertia*. It goes something like this:

Things which can move tend to stay still and it takes a force to start them moving.

Once they are moving they tend to go on in a straight line at a steady speed. It takes a force to speed them up, slow them down, or change their direction.

This tendency of things to 'go on as they are' is called *inertia*. The word comes from a Latin word meaning 'laziness'.

You can use the inertia idea to explain what is happening in Fig. 7.4. The horse stops. Because the rider has inertia she tends to keep going in a straight line at a steady speed. The force of gravity pulls her down. The more inertia she has, the greater her tendency to keep going.

Fig. 7.4 The rider's inertia keeps her going in a straight line at a steady speed

Can you use the inertia idea to solve this puzzle?

An ice hockey puck (see page 130) is being swung round and round on the end of a string on an ice rink. Suppose the string breaks when the puck is at point X (Fig. 7.5). Which of the dots numbered 1–9 will the puck hit? (The answer is on page 161.)

Because of the way science developed we have two names for inertia. When we set out to measure how much inertia something has – how hard it is to start it moving, or speed it up, or change its direction – we stop talking about inertia and start talking about *mass*. In other words, *mass is the name we give to inertia when we measure it*. It seems silly to have two names for the same thing, but scientists have learned to live with it. At first it is a nuisance to remind yourself that when you are talking about the mass of something you are talking about its inertia – but you soon get used to it!

About 100 years ago, scientists from all over the world got together at a special conference in France. They decided to keep an international 'standard' of mass. The masses of everything else in the world – even the planet Earth itself – would be compared with this special object.

So, in a special room at the International Bureau of Weights and Measures at Sèvres near Paris there is a small gold-coloured cylinder. It is made of an alloy of the metals platinum and iridium. The cylinder is called *The International Prototype Kilogram*. Most of the countries of the world have copies of it.

Suppose your mass is 40 kg. That means that your inertia is 40 times greater than the golden cylinder at Sèvres. And that means that you are 40 times harder to start moving than the cylinder. Once you are moving you are 40 times harder to speed up, slow down, or change in direction.

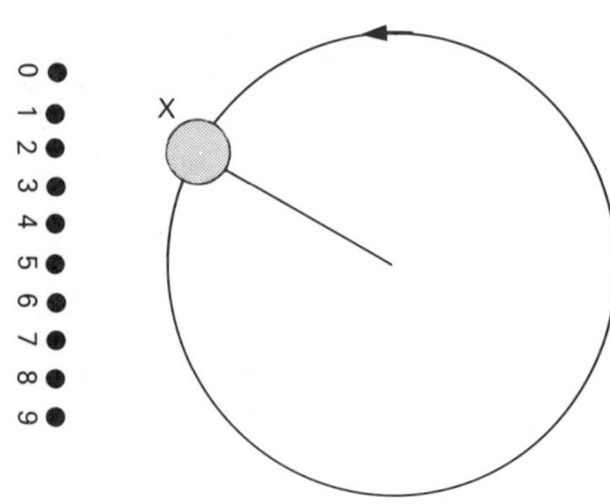

Fig. 7.5 Which of the dots will the puck hit if the string breaks when the puck is at point x?

7.3 Pushes and pulls

Figure 7.6 shows different forces in action. Look at each picture in turn. Try to say what the force – the push or the pull – is doing. For instance, picture (a) shows a golf ball being struck by a club. You might say: 'Here the force

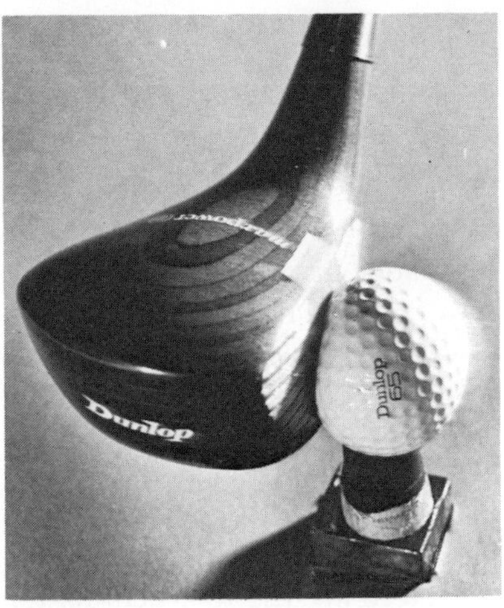

(a)

(b)

(c)

(d)

Fig. 7.6 (a) A golf club hitting a ball
      (b) A skier banking on a turn
      (c) A cyclist making an emergency stop
      (d) A rocket lifting off

is causing a change in shape'. Of course, a split second later the push from the golf club will have another effect – the ball will start moving – but in the picture, change in shape is the main effect.

Scientists have a special way of talking about forces in action. They use the words *apply*, *applying*, and *applied*. For example, a scientist would say that the golf club is applying a force which is causing the ball to change shape.

IN YOUR NOTEBOOK

✳ Write a heading: *Forces in action*

✳ Copy this table and try to fill in the gaps:

Event	Force applied by	Effect of force
Golf club strikes ball	Golf club	Change in shape of ball
Cyclist puts on brakes	Brake blocks	
Rocket engines fire	Expanding gases	
Skier banks over	Skis and sticks	

* Write this important statement:

  Forces can do two things – they can have two effects. Forces can use things to change shape and forces can overcome the inertia of things. That is, they can start things moving, speed them up, slow them down, change their direction, and stop them.

* Try to find some pictures of forces in action to stick in your notebook. Make a note to say which of the effects is being shown in each picture.

7.4 Make a push-pull forcemeter

On page 13 you read about forcemeters. Perhaps you made one of your own for measuring small forces – small *pulls* that is. Here are the instructions for making a more useful one which will measure both pushes *and* pulls. Fig. 7.7 will help.

Investigation 7.1 Making a forcemeter and measuring pushes and pulls

You will need:
a 30 cm length of broom handle
a 15 cm length of cardboard or plastic 'mailing tube' 3–4 cm diameter (this is a
  tube sold by stationers for sending posters and pictures through the mail)
a small cup hook
2 thick rubber bands – 10 cm × 6 mm is a good size
a strip of 'write-on' sticky tape
a single hole-punch
a 0–30 N forcemeter from the science lab

You read about calibrating an instrument on page 14.
  When you have calibrated your forcemeter, use it to measure some everyday forces such as the force to . . .

pull a rubber sucker off a wall;
open a ring pull can;
pull the plug out of a sink (try it with the sink empty then full of water).

  Measure the force needed to start a brick moving along the floor, first pushing then pulling. Is there a difference? Then make a Meccano trolley for the brick and measure the starting force again. Put different sized wheels on the trolley – do they make a difference?

IN YOUR NOTEBOOK

* Draw a diagram of your push-pull forcemeter.

* Write about the measurements you have made and any things you have found out.

7.5 The sticky force

Instead of a ball, ice hockey is played with a thick disc of hard rubber called a 'puck'. If you have watched ice hockey on TV you will have seen the puck

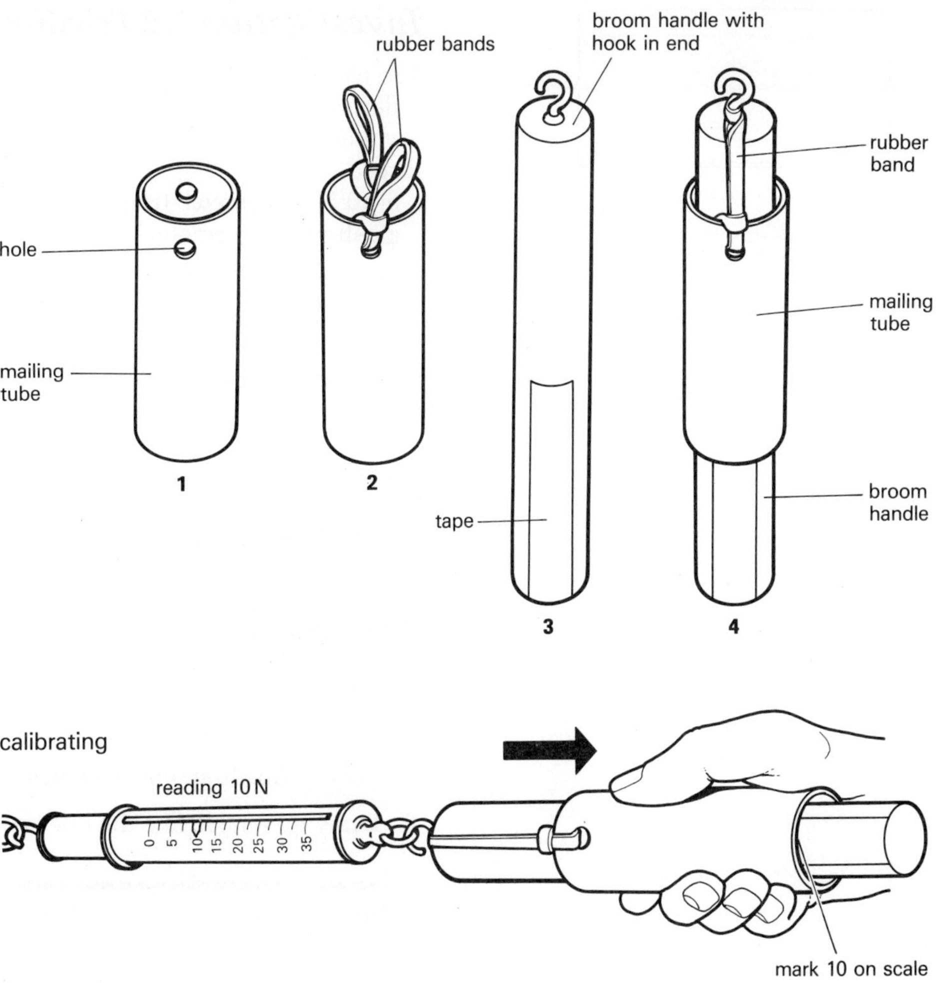

Fig. 7.7 Investigation 7.1

slide from one end of the rink to the other. But suppose that you wanted to play hockey with a puck in the school playground in the summer! What problems would you have?

Of course it would be silly to try it! You would have to hit the puck very hard to make it slide over the ground at all – and it would not get very far before it stopped.

The reason is because there is much more *friction* between the puck and the playground than between the puck and the ice. But what does that mean? What causes friction?

You can investigate friction forces in the science lab.

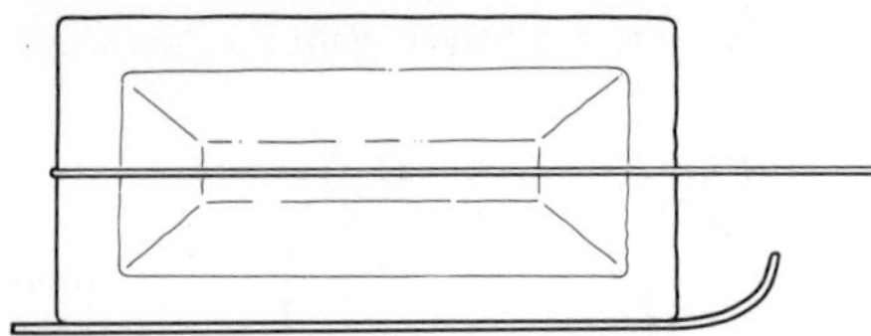

Fig. 7.8

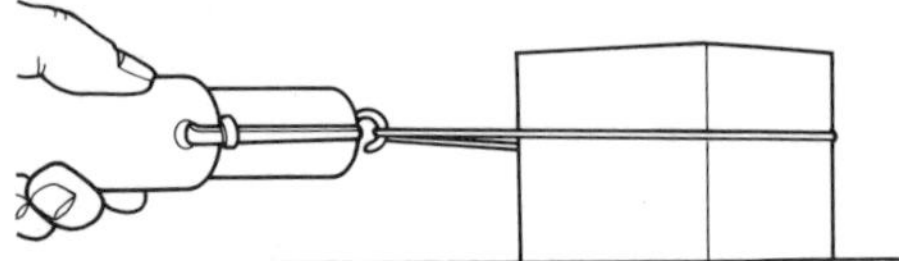

Fig. 7.9 Investigation 7.2

Investigation 7.2 Friction forces

You will need:

a flat metal tray

several steel blocks, 5 cm × 4 cm × 3 cm*

thread

a 0–10 N forcemeter from the lab – or your own push-pull forcemeter

sandpaper

talc

polystyrene beads

a dropper

can of light oil

1 Put one block on the tray with its largest face down. Measure the force you need to start the block moving – the force that is just big enough to overcome the friction force. Pull very gently, watching your forcemeter until the block just starts to slip. Jot the force down.

2 Measure the force you need to keep the block sliding smoothly and steadily along. Ask a friend to read the forcemeter whilst you concentrate on pulling. Do several runs and jot the forces down.

3 Put a large sheet of sandpaper on the tray and do parts 1 and 2 again. Try some other surfaces – take away the sandpaper and sprinkle talc or polystyrene beads on the tray, or put a few drops of oil on it.

4 Put a second block on top of the first one, and do one or two of your investigations again. Does the extra weight make a difference?

IN YOUR NOTEBOOK

✷ Write a heading: *Investigating friction*

✷ Draw up a table like the one below and fill in your results.

Investigation	Force needed to overcome friction				
	without anything	on sandpaper	with talc	with beads	with oil
Large face of block: starting					
Large face of block: sliding					
Two blocks: starting					
Two blocks: sliding					

5 Stand one block on its end (on its smallest face) and measure the starting and sliding friction again. Are the forces different with a smaller 'area of contact'?

Scientists have carried out many investigations of friction like yours. They have repeated their measurements again and again until they are quite sure of the patterns.

They have found that *friction depends on the kinds of surfaces, and it depends on the force pressing the two surfaces together.*

* If you are short of blocks you can use bricks on metal sheet 'sledges'.

You probably found that the friction was greater with two blocks piled up than with one.

The friction *does not seem to be affected by* the area of contact, however.

How can these things be explained? Think again about the ice hockey puck . . .

If you were to slice the puck in two from top to bottom and use a microscope to look edge on at its 'flat' bottom you would find that it was anything but flat! Fig. 7.10 shows the surface of a highly polished steel needle as seen through a powerful microscope. Under higher magnification it would seem even rougher – full of lumps and hollows a few tens of atoms across.

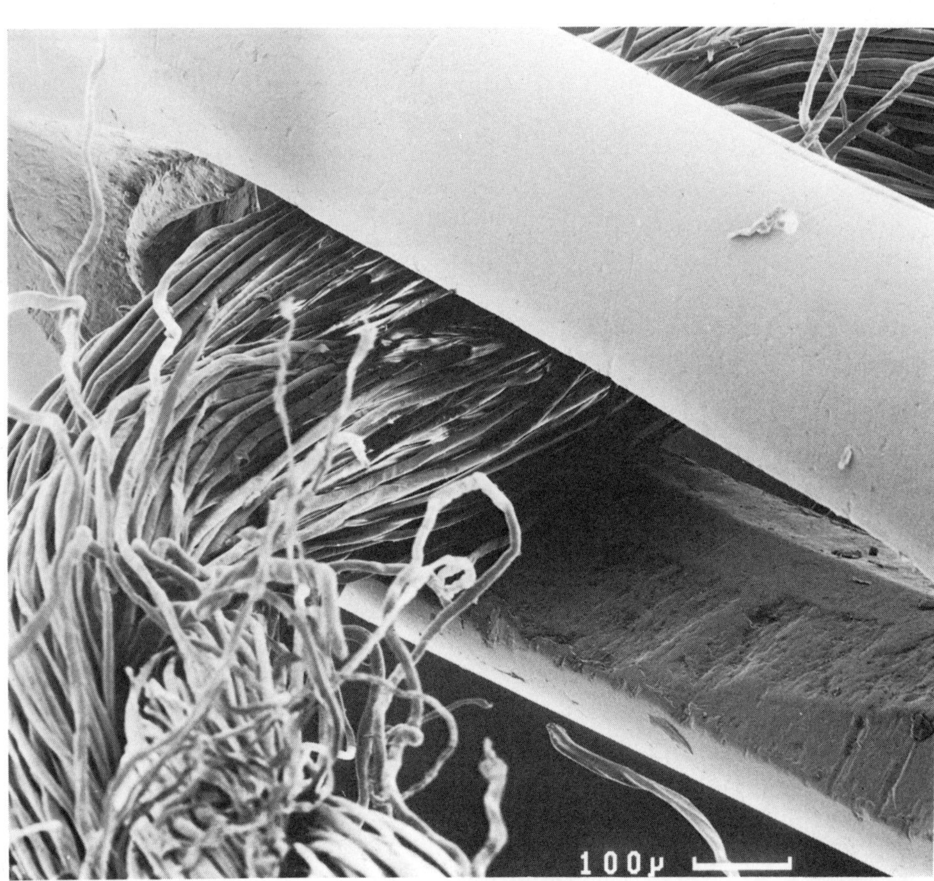
Fig. 7.10 Scanning electron micrograph of the eye of a needle

When two surfaces rub together, the tiny humps and hollows interlock. Many of the humps on both sides join together as if the atoms in them were part of the same piece of material.

This is what would happen with the ice hockey puck as it slid over a playground. The tiny points on its bottom would join up with those on the playground, and then would be torn apart as it slid along. Of course that would soon slow it down. The force caused by this sticking and breaking is what we call friction.

If you used a puck which was twice as heavy, many more humps and hollows would join up so the friction would be greater.

There is far less friction on ice because the puck slides on a thin cushion of water.

If a puck was squashed so that it had a much bigger area, the friction would be unchanged. Can you explain why the 'area of contact' does not make much difference to the friction? Think about it for a while and make a few notes. Then talk about it with your teacher.

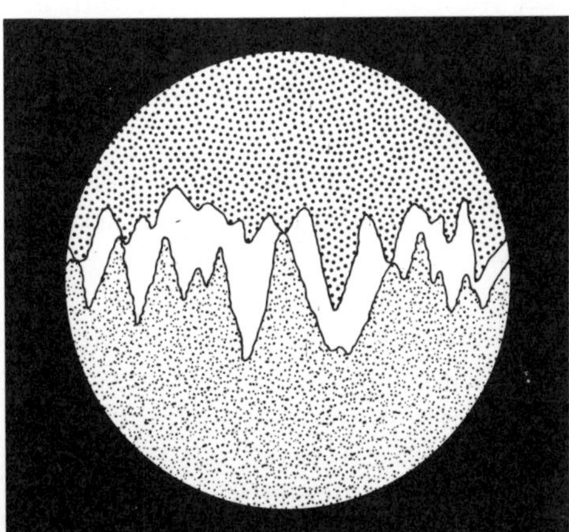
Fig. 7.11 Friction is caused by the tiny points on the bottom of one object joining up with those on the top of another object over which it is moving

7.6 Friction – friend as well as foe

People tend to think of friction as only a nuisance. It wastes fuel in getting engines started and in keeping them running. Moving parts have to be separated by keeping them well oiled. Engines, drive bands, and tyres wear out.

But what would life be like without friction? Unless we could grab hold of something it would be very difficult to get moving. If we did get moving it would be hard to stop or change direction. Shoe laces would unfasten themselves, nails and screws would fall out of walls, you would not even be able to strike a match!

IN YOUR NOTEBOOK

✻ Write a short story about a scientist who invents the perfect lubricant – a kind of oil which completely prevents friction whenever it is used. Perhaps he or she accidentally spills some on the floor of the lab . . .

7.7 The universal force

Fig. 7.12

There is a famous story about Isaac Newton and an apple. You may have heard it.

In the year 1666, the Great Plague was killing thousands of people in London. Sensibly, Isaac Newton had gone to stay with his mother at her farm in Leicestershire. One day, sitting in the garden, he saw an apple fall to the ground.

The way the story is usually told, he is supposed to have realised that the apple was falling because the Earth was pulling it down. But that seems unlikely – people had been talking about the force of gravity for years. What is more likely is that the apple triggered off a sudden flash of inspiration in Isaac's mind – great scientific ideas often come like that.

Suddenly, he saw how to solve a problem which had worried him for weeks – how to explain how the Moon stayed in its orbit round the Earth, and how the planets stayed in their orbits round the Sun. He realised three important things:

1 The Earth was attracting the apple but the apple was also attracting the Earth!
2 Everything in the universe – the apple, the Earth, the Moon, the Sun and the planets – attracts everything else.
3 The more massive the object, the more strongly it attracts things towards it.

The Earth is enormously massive – about 6 million million million million kilograms – so it attracts things very strongly to it. The apple's pull is so weak that it has no real effect on the Earth – but its pull is still there.

Nowadays, any scientist will tell you that the Moon stays in its orbit because it is attracted by the Earth. Without the pull of the Earth, the Moon would go off in a straight line because of its inertia. The force of gravity acts like an invisible string pulling it inwards just enough to keep it moving round. The Sun is so massive that its gravity pull can reach out to keep the giant planet Jupiter in its orbit more than 700 million kilometres away.

You can measure the Earth's pull on your body. Just stand on a set of bathroom scales calibrated in newtons. As you stand there, say to yourself: 'The Earth is pulling me downwards towards its centre with a force of _______ newtons. I am feeling the same force that keeps the moon in its orbit'.

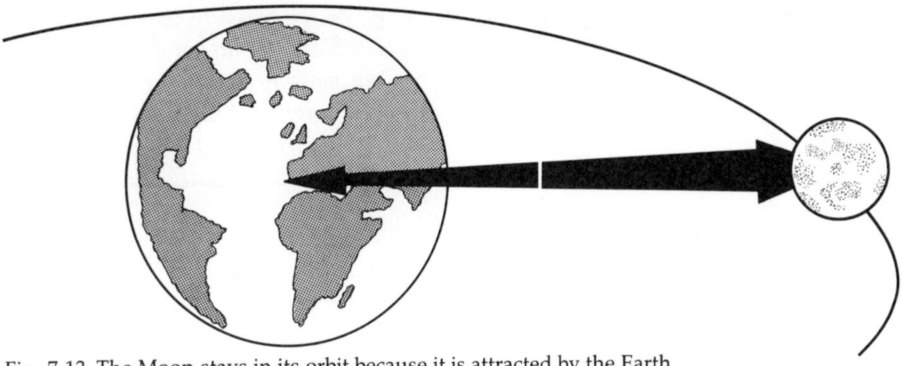

Fig. 7.13 The Moon stays in its orbit because it is attracted by the Earth

The Earth's pull on your body is called your *weight* of course.

How much does the Earth pull on a 1 kg mass? You can measure it. Just hang a l kg mass on a forcemeter (0–15 N or 0–20 N). You should get a value of about 10 N.

The weight of a kilogram is very nearly the same anywhere on the Earth but it is a different story elsewhere in the universe. For instance, if the astronauts had done your experiment on the Moon the kilogram would have weighed only about $1\frac{1}{2}$ newtons (about a sixth of its weight on Earth).

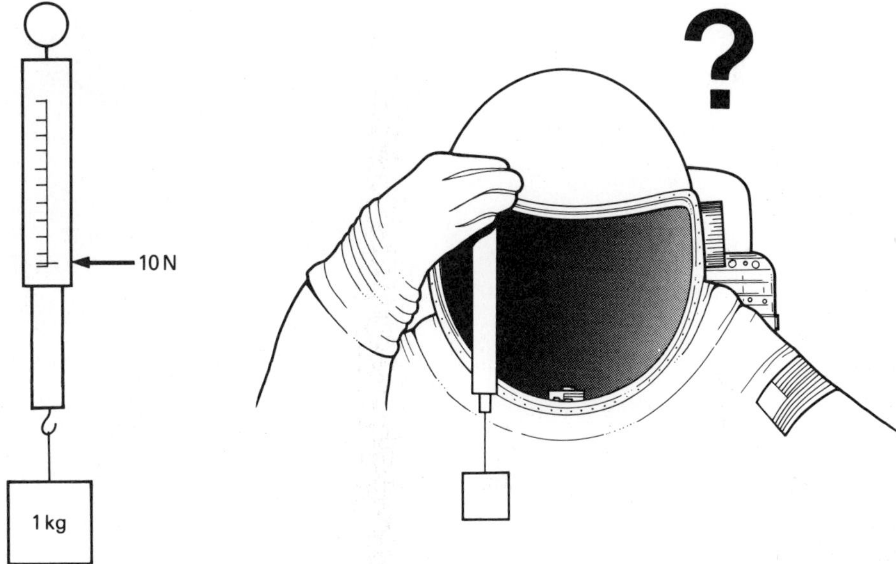

Fig. 7.14 The forcemeter on the left shows the weight of a kilogram on the Earth. What is the weight of a kilogram on the Moon?

Also the astronauts had only one sixth of their Earth-weight. For this reason they had to be careful not to go bounding along – risking damage to their pressure suits.

Now that you know the weight of a kilogram and your own weight on Earth you can work backwards to find your own *mass* in kilograms.

For example, if you know that your weight is 400 N you can say that your mass is very nearly 40 kg.

IN YOUR NOTEBOOK

✴ Write a heading: *Inertia, mass, and weight*

✴ Write these statements, filling in the gaps:
Inertia means the tendency of things to stay still or move in a straight line at a
_____.

(Continued on next page)

When we measure inertia we call it _______. It is measured in kilograms.
The Earth's pull on a mass of 1 kg is about _______ N.
My mass is _______ kg.
My weight is _______ N.
On the Moon my mass would be _______ kg and my weight would be _______ N.

7.8 Force multipliers

To a scientist, a *machine* is a 'device for applying forces'. Almost all of them are 'force enlargers' or 'force multipliers'. You can apply a much larger force by using them than you could with your bare hands.

Fig. 7.15 shows a collection of simple machines. Can you say what they are designed to do?

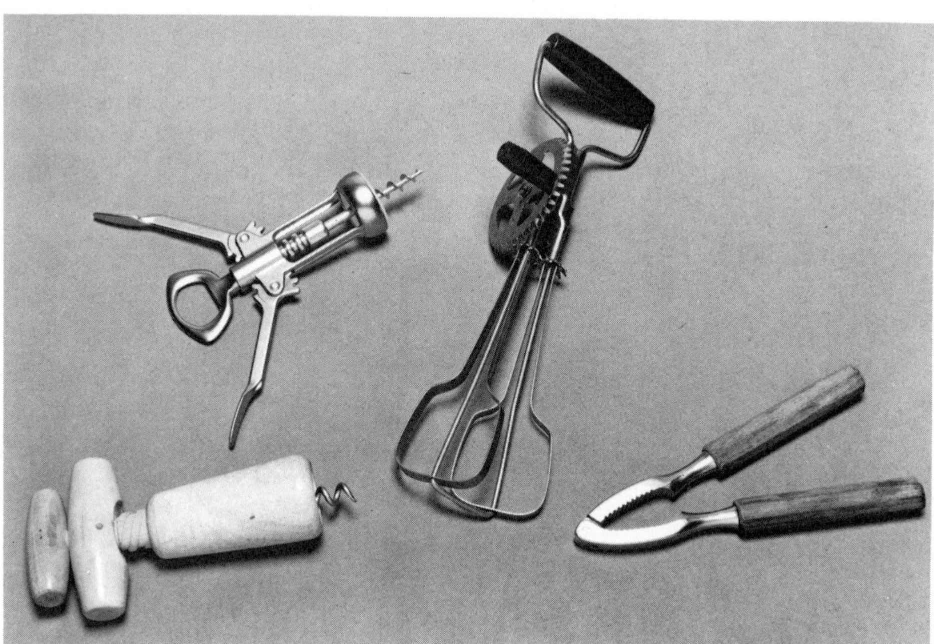

Fig. 7.15 What are these simple machines designed to do and how do they do it?

If you apply quite a small force to the corkscrews or the nutcrackers, they will apply a much larger force for you. Enough force in fact to get the cork out of a wine bottle or to crack a nut.

The egg whisk is different. The force it applies for you is actually *much less* than the force you apply to the handle. But the beaters move much further than your hand in the same time – they move much faster. So the egg whisk is a *distance multiplying* machine, or a *speed multiplier*.

A few machines are like the egg whisk; but *you cannot have a machine that multiplies both forces and distances.*

Fig. 7.16 shows a machine with a very big 'force multiplying ability'. You apply a force of 50 N or so to the screw jack and it applies a force of hundreds of newtons for you – enough to lift a car!

Scientists call the force which you apply to the machine, the *effort force* or just the *effort*. The force the machine applies for you is called the *load*.

The advantage of using a machine like the screw jack is obvious. Unless you were very strong, you could not lift a car without one. But there are disadvantages to using machines too. You will know what they are when you have carried out the next investigation.

Fig. 7.16 A screw jack

7.9 Investigating a pulley system

Car accessory shops sell 'midget hoists' which can be used to do heavy lifting jobs. There may be one of these useful machines in your science lab.

Look at the hoist closely and you will see that it has two main parts. Each is a sort of frame holding several small grooved wheels. These grooved wheels are called *pulleys* so the hoist is called a *pulley system*.

The two frames are linked together by a cord which loops round and round the pulleys. Fig. 7.18 shows how the cord goes round. To make it clear, the frame has been 'stretched' by the artist.

Notice that the cords go over the tops of the top set of pulleys and underneath the bottoms of the bottom set. When you pull on the effort cord you unwind some of the cord from the system. All the cords between the pulleys get shorter, and the bottom frame is lifted up.

How many cords run between the two frames (not counting the effort cord)? Count them.

There are eight cords pulling upwards on the bottom frame. So if your effort force was 100 N you might expect the bottom frame to be pulled upwards with a force of 800 N. Find out whether it is . . .

Fig. 7.17 A hoist being used to lift an engine from a car

Investigation 7.3 Finding the 'force magnification' of a pulley system

1 Set up the hoist as shown in Fig. 7.19. *Check with your teacher that the top set of pulleys is securely fixed to a strong part of the ceiling. (Many labs have a beam in the ceiling for this.)*

2 When you are ready, pull on the effort cord using a forcemeter. Pull until the stool and its occupant are just lifted off the ground. Get someone to read the forcemeter.

3 Measure the total load force. That is, everything that all the strings between the pulleys have to lift: the stool, its occupant, the cradle of ropes, and the bottom frame of pulleys. Do this by asking the person to stand on a set of newton bathroom scales whilst he or she holds up the stool and cradle. (The frame and pulleys are very light compared with the rest of the load so you can forget about them if you like.)

4 Compare the effort with the load. How much does this machine magnify forces? Is the load eight times greater than the effort? Is it more or is it less?

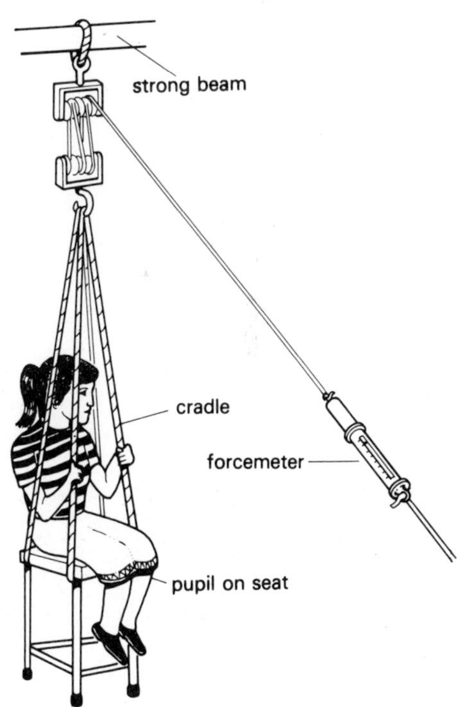

Fig. 7.19 Investigation 7.3

Investigation 7.4 Measuring the distances moved in a pulley system

You will need a metre scale, a metric tape measure, and several people to help with this.

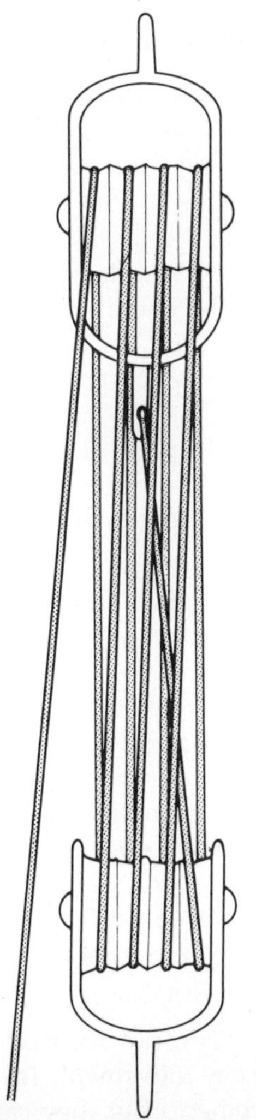
Fig. 7.18 The stringing of pulleys in a hoist

1 Set up the apparatus again. With the stool resting on the ground 'take the strain' holding the effort cord taut in your hands. Then move away from the pulley system pulling on the cord. Get someone to call out when the stool and its occupant have risen 20 cm. *Slowly* let them down again.
2 Do the same thing again, but this time measure how far you move the effort cord as the load rises 20 cm. Perhaps you could use two tall stands to mark the places where your hands were when you started and stopped pulling.

IN YOUR NOTEBOOK

✳ Write a heading: *Investigating a machine*

✳ Draw a diagram showing the midget hoist in action.

✳ Draw up a table like the one below and fill in your results:

Load force (N)	Effort force (N)	Distance moved by load (m)	Distance moved by effort (m)
		0.2†	

† 20 cm or 1/5 of a metre.

✳ Write a few notes to explain what a machine is and what machines do.

7.10 Move-ability

The word 'energy' was once used only by scientists. Nowadays, everybody uses it. People talk about 'the energy crunch' and 'the energy crisis' – we know that the Earth's supplies of energy are running out. In 1979 and 1980 the Government organised the 'Save it' campaign to try to persuade people to 'use less energy'.

But what is energy? What does the words mean? In particular, what does it mean to a scientist?

A good way to find the answer is to look at some things which you know have energy. For instance . . .

1 *Foods have energy*. You get your energy from the foods that you eat (with the help of the oxygen gas that you breathe in). What does that mean? What does your food help you to do?
2 *Fuels have energy*. A rocket gets its energy from the fuels it burns. So does a jet plane or car. What does the fuel give to the rocket? What difference does it make? What does it help the rocket to do?
3 *Electric currents have energy*. An electric mixer gets its energy from the electric current pushed through it by the mains. What does that mean? What does the current give to the mixer? What happens when the current is switched off – when the supply of energy is taken away?
4 *All moving things have energy*. The moving wind gives energy to a windmill. What does that mean? What happens if the wind stops?

You know that all of the examples involve energy. You can see that they all involve movement. That *does not* mean that energy *is* movement. It does not make sense to say that foods and fuels have movement. But they can help to *cause* movement, and that is one of the secrets of energy.

Fig. 7.20 Where do these machines get their energy from?

Putting it in a straightforward way:

energy is the ability to cause movement or to cause change in movement.

So anything that can cause movement or change in movement has energy. Whenever you see movement you know that energy is involved.

You may have seen an apparatus like the one in Fig. 7.21. A lamp shines on to a silicon solar cell (as used to generate electricity for satellites and spacecraft). The solar cell is connected to a small electric motor.

The motor starts as soon as the lamp is switched on. It stops when the lamp is switched off. You can say that the light from the lamp has *caused* the movement of the motor. So the light must have energy. This is true for all kinds of light – even the 'invisible' kinds such as ultra-violet, infra-red, X-rays, and radio waves.

It seems a very simple idea but scientists have found it enormously useful in organising their ideas.

When you talk about energy you need to keep reminding yourself that it is an *ability* – 'movement-causing ability' if you like. It is *not* a thing or a kind of stuff.

IN YOUR NOTEBOOK

✳ Write a heading: *Energy*

✳ Write this important definition:

  Energy is the ability to cause movement or to cause change in movement.

✳ Make a list of the things mentioned on these pages which have energy. Draw some pictures to illustrate your list.

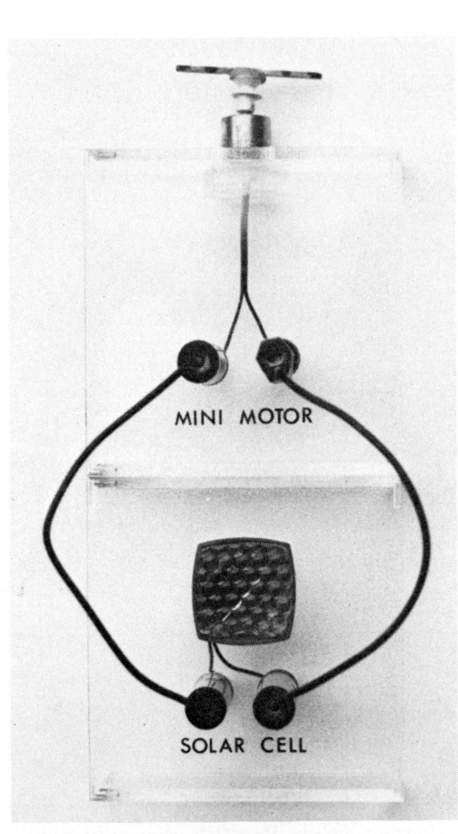

Fig. 7.21 Silicon solar cell apparatus

7.11 Transferring energy

We humans have become experts at inventing devices for passing on energy from one thing to another – for transferring energy. By transferring energy the devices are able to do useful jobs.

For example, bows and arrows were being used centuries before the birth of Christ. A bow and arrow transfers energy from food through a person's muscles to an arrow.

We can draw an *energy transfer diagram* for the bow and arrow. Notice that the things named in the boxes are the things which have the energy in turn. So the energy (ability to cause or change movement) starts with the food and finishes with the arrow.

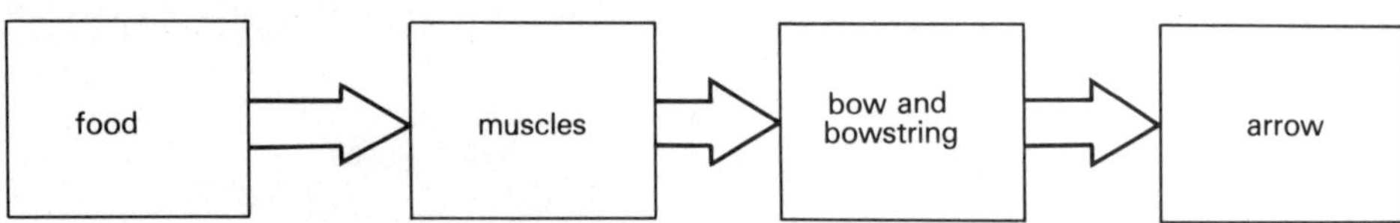

Fig. 7.22 Energy transfer diagram for bow and arrow

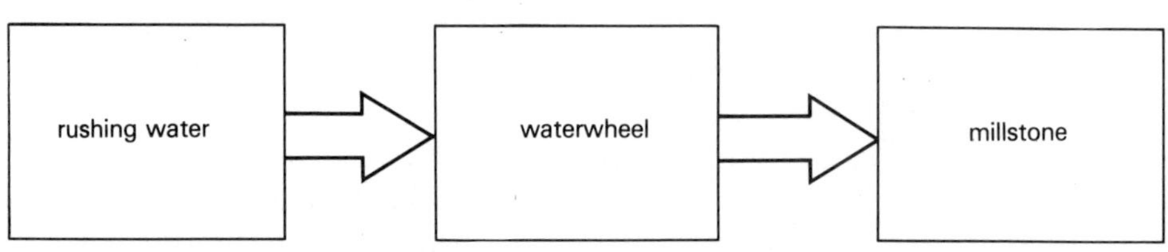

Fig. 7.23 A water mill. This one is Ifield Mill, near Crawley, Sussex. There has been a water mill here since the 16th century.

In the year 1086 there were about 6000 working water mills in England. Even in 1830 a third of England's cotton mills were still using water wheels as their main energy transfer device.

The energy transfer diagram shows that energy is transferred from fast-moving water to the waterwheel then to the millstone (through driving shafts and belts).

The first fully working steam engine was built in 1712. The first steam engines were used to do the same jobs as water wheels. It was 90 years before tracks were used to transfer energy to the body of a self-moving engine – a 'locomotive' – and the trucks it pulled. The steam engine transfers energy from wood, coal or some other fuel to superhot steam in its boiler, then to its driving wheels.

Fig. 7.24 Stephenson's Rocket

Not all energy transfers involve movement. For instance, a *laser* (invented around 1960) transfers energy from an electric current to a very strong beam of light. You do not *have* to see movement to know that energy is being transferred. Electric current goes into the laser and light comes out. You know that both these things have energy so it must be transferred.

IN YOUR NOTEBOOK

The table below is a timeline of energy transfer devices. Copy it and try to fill in the gaps. Choose the words you need from this list:

Moving fluid (liquid or gas)   Electric current
Moving object   Light
Spinning driveshaft   Sound waves
Hot gas   Chemical substances

One or two of the gaps have been filled to give you a start.

Year of first practical version	Transfer device	Transfers energy . . .	
		From	To
1800	Electric battery	Chemical substances	Electric current
1820	Camera and film		
1831	Generator (dynamo)	Spinning drive shaft	
1833	Water turbine	Moving fluid (water)	Spinning drive shaft
1840	Bicycle		
1860	Internal combustion engine*		
1876	Microphone	'Sound waves'	
1878	Electric light bulb		
1896	Radio set		
1902	Photocell	Light	
1926	Rocket engine		Moving object
1939	Jet engine		

* Motor car engine

7.12 Uphill energy

Nowadays, most people when they hear the word 'energy' think of coal, oil, natural gas or electricity. That is because they have read about energy in connection with North Sea oil and gas, and the 'Save it' campaign. It can come as quite a surprise to hear that the peaceful water in high reservoirs like the one in Fig. 7.26 has enormous amounts of energy.

Fig. 7.25 A laser being used to cut into metal. The beam itself is in the tube coming down from the top right. The bright lines coming out from the metal are the paths of white-hot metal particles. Energy has been transferred to them from the laser.

Fig. 7.26 A high reservoir such as this has enormous amounts of energy

Suppose you were talking to someone about the water in the reservoir. How could you convince him or her that the water had energy? Remember – energy is the ability to cause movement or to cause change in movement. So, the problem is to show that the water in the reservoir can cause movement.

You might explain about the hydro-electric power stations which are fed by the water from some reservoirs. You could say that if the water is let go it will rush downwards. It can be made to turn a water turbine. The turbine can drive a generator which will push electricity through wires to our homes.

The energy transfer diagram looks like Fig. 7.28.

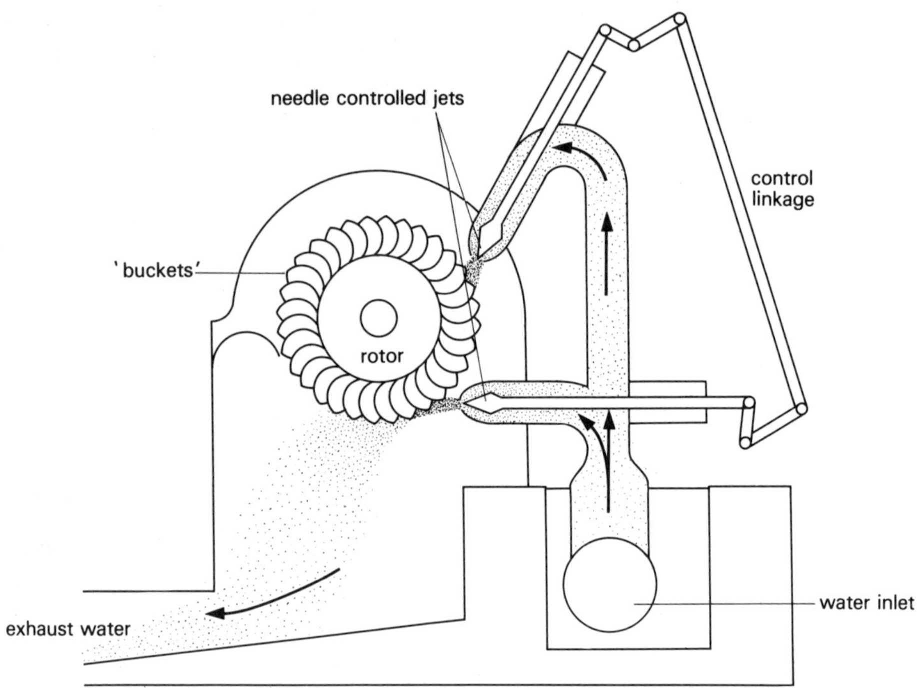

Fig. 7.27 Water turbine. This type is called a Pelton wheel. Which way round would the rotor go?

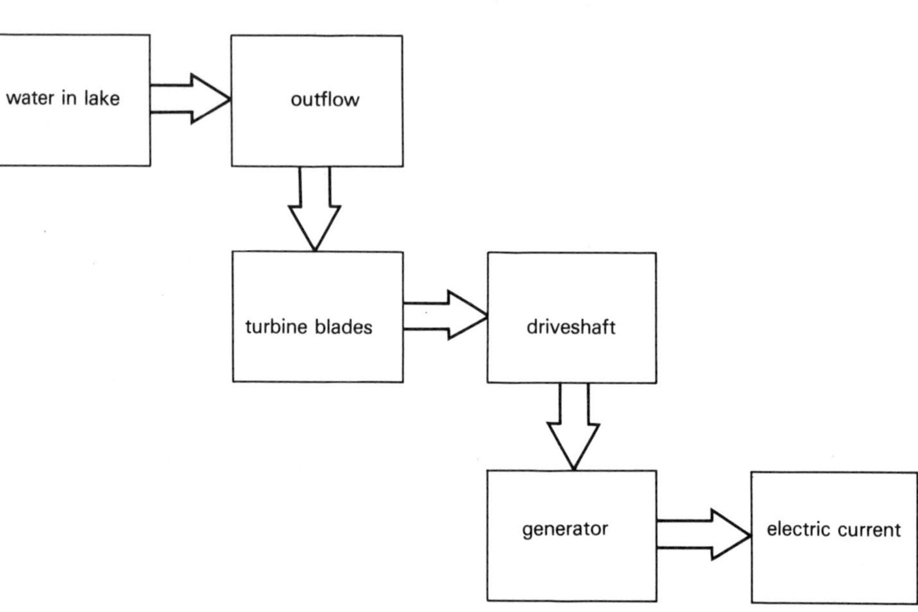

Fig. 7.28 Energy transfer diagram for a hydro-electric power station

Anything which is high up has more energy than something which is lower down. The higher up it is, the more energy it has. For example, a book on a shelf has energy. You could, if you wanted to, rig up a system of strings and pulley wheels so that, if you pushed the book off the shelf it would make something move! Perhaps you could make it turn a small windmill or lift up a small object.

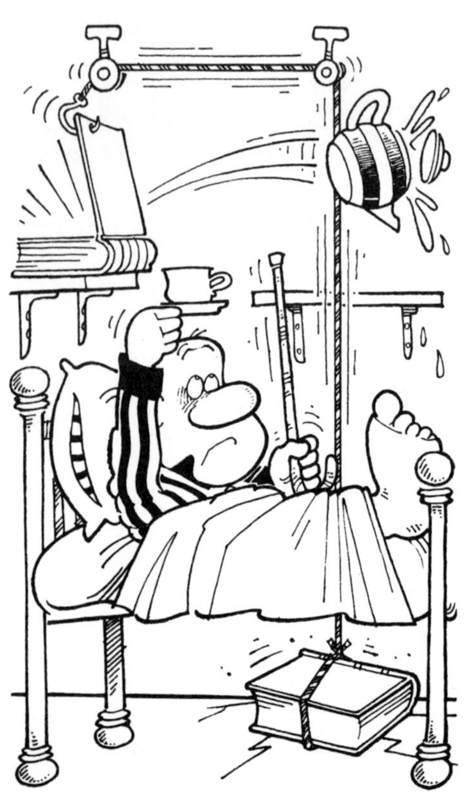

Fig. 7.29

Many years ago, scientists used this idea of 'uphill' energy to work out a way of *measuring* amounts of energy transferred. They were particularly interested in measuring how much energy people transferred whilst they were working. They invented stories like the one in the puzzle below. Can you solve the puzzle?

7.13 The box and shelf puzzle

Three people are working in a store-room. They have 9 equally heavy parcels to put away on a stack of 6 shelves.

The first person picks up 2 parcels, climbs up the ladder and puts them on the top shelf (shelf 6).

The second person thinks for a while, then climbs up and puts 4 parcels on shelf 3. She says: 'There – now I have done as much work on the parcels as you'.

The third person has 3 parcels left. The others insist that he 'does just as much work' as they have. What shelf must he put the three parcels on? Jot your answer down.

You could draw up a table showing the jobs done by the three people. It would look like this:

Number of parcels	Height of shelf	Amount of work done
2	6	
4	3	
3	?	

What number do you put instead of the question mark? (Answer is on page 161.)

Suppose someone else came along with 12 more parcels. Which shelf should he or she put them on to do the same amount of work as the other three people?

All four people will agree that they have done the same amount of work in moving the parcels. They will all have transferred the same amount of energy from their food to the parcels.

What can you do with the numbers in the first two columns of the table so that you get the same number for each person in the last column?

Of course, the answer is 'multiply them together'

$$2 \times 6 = 12 \quad 4 \times 3 = 12 \quad 3 \times 4 = 12$$

and for the fourth person, $12 \times 1 = 12$.

This method for getting a number for the amount of energy transferred is fine when you are talking about 'standard' parcels which are all of the same weight. But what if the parcels have different weights?

Thinking along these lines, scientists chose to measure the energy transferred when this kind of work is done by taking the total weight of the parcels and multiplying it by the distance moved. Finally, they arrived at this formula:

> Energy transferred = Force applied × Distance moved

If the force is measured in newtons and the distance is measured in metres the energy units are called joules (J).

The name joule was chosen to honour the English scientist James Prescott

Joule who worked about 140 years ago and made many investigations of energy transfers.

In the history of science, the connection between energy transfers and the work which people do has been so strong that a force multiplied by a distance is often called an amount of *work*.

IN YOUR NOTEBOOK

✳ Write a heading: *Measuring energy transferred*

✳ Write these statements:

You can calculate the energy transferred to 'uphill things' and moving things. You multiply the applied force by the distance moved. The energy is measured in joules (J).

An amount of energy measured like this is called an amount of work. So:

  force × distance moved = work (J)

✳ Try these problems to do with the box and shelf puzzle. (Answers are on page 161.)
Imagine that each parcel weighs 50 N and the shelves are ½ a metre apart.

  1 How much energy would you transfer to one parcel in lifting it on to the bottom shelf?
  2 Suppose you put two parcels on shelf 6. How much energy would you transfer to *each* of them?
  3 Suppose you moved a parcel from the top shelf (shelf 6) to shelf 1. How much less energy would it have?
  4 Where would you put three parcels to give them a total of 150 J of energy? (Which shelf?)
  5 Think of two different ways of transferring 75 J of energy by putting parcels on shelves.
  6 The *total* amount of energy transferred by the first person in the puzzle (the one with 2 parcels) was actually much greater than that transferred by the person with 12 parcels. Can you explain why?

7.14 Working against the clock

It is interesting to measure the amounts of energy you can transfer using different sets of muscles in different activities.

Figs 7.30, 7.31, 7.32 and 7.33 show four different activities you could try. Ask you teacher to arrange them as a 'circus' (or a 'circuit') so that you and your classmates can go round doing each one in turn.

Notice that each activity lasts for *20 seconds*. You will need to work with a friend so that you can keep to time. Measure the distances in *metres* (1 cm = 0.01 m).

You could add other activities to the circus such as:

cycling on a cycle ergometer;
using a 'chest expander';
opening a self-closing door;
pushing an iron across an ironing board;
pulling a loaded trolley along the playground.

You must be certain that you are fit and healthy before you try any of these activities.

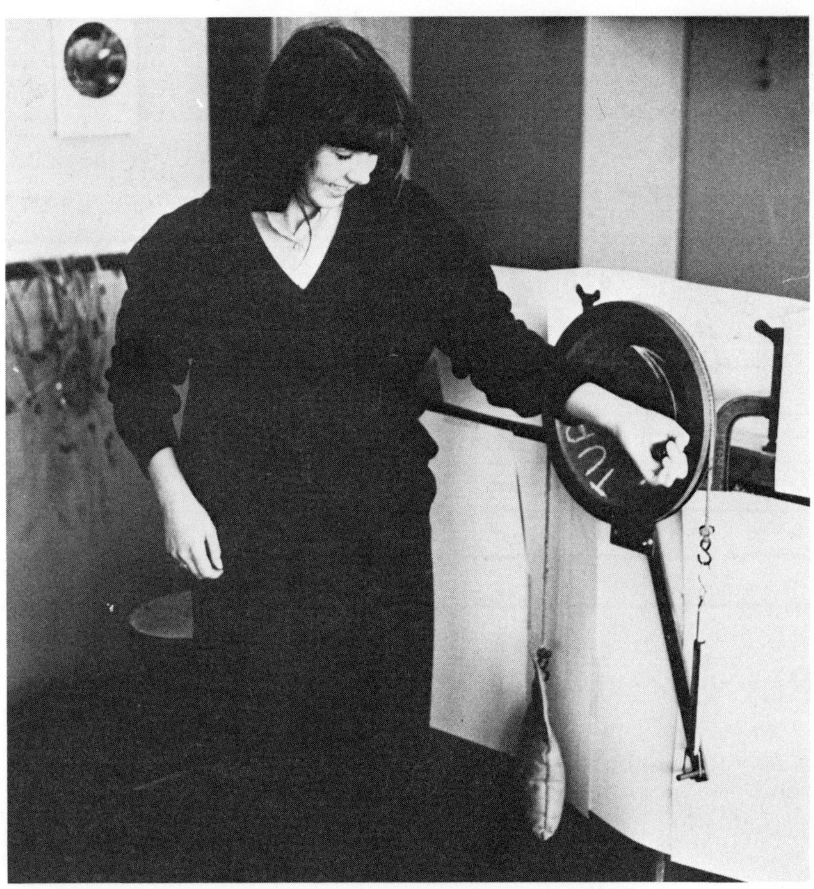

Fig. 7.30 Hand ergometer – how many turns in 20 seconds?
Force = change in forcemeter reading caused by turning the wheel
Distance = circumference of wheel × number of turns

Fig. 7.31 Step-ups – how many steps in 20 seconds?
Say to yourself: 'Left up, right up, left down . . .' as you
go. Count every left step up.
Force = your weight
Distance = height of box × number of left step-ups

Fig. 7.32 Press ups – how many in 20 seconds?
Keep your back straight and lift your whole body off the ground.
Force = your arm force (do one press-up on a set of newton bathroom scales
before you start)
Distance = distance you lift your shoulders each time × number of times

Fig. 7.33 Sandbag lifting – how many on to the bench in 20
seconds? Put all the sandbags flat on the floor to begin with.
Force = total weight of all the bags you lift
Distance = height of bench top above floor

IN YOUR NOTEBOOK

✳ Write a heading: *Measuring my work*

✳ Draw up a table like the one below and fill in your results. Leave the right hand column empty for the time being.

Activity	Force (N) F	Distance (m) s	Work done in 20 seconds (J) $(F \times s)$	Power

7.15 Power!

In dragster racing the aim is to reach the highest possible speed before you get to the end of a measured length of track (called a drag strip).

Fig. 7.34 Dragster racing just after a start

Suppose you were comparing engines for a dragster. You would want to know *how quickly* each engine could transfer energy from the fuel to its road wheels and body. That is, how much energy it could transfer *in a second*.

For most energy transfer devices, it is not the amount of energy·they can transfer that matters, but how quickly they can transfer it.

Scientists have a special name for *the energy transferred by a device in one second*. They call it the *power* of the device.

Manufacturers of energy transfer devices usually label each one with its power. For example, some lawn mowers are labelled 900 watt. That means they can transfer 900 joules of energy every second.

Joules per second are called watts to honour one of the great designers of engines – James Watt, a Scottish engineer who worked about 200 years ago.

Electric light bulbs are labelled 40 W, 60 W, and so on. A 60 W bulb can transfer 60 joules of energy from the mains every second.

7.16 Calculating power

Suppose you want to measure the power output of a model steam engine, or an electric motor, or your own body. What do you need to know? How do you work it out?

You can go about it like this:

1 Find the force in newtons which the device applies when it is working normally. (Call this F.)
2 Find the distance in metres moved during a test run. (Scientists usually call this s.)
3 Find how long in seconds the test run takes. (Call it t.)
4 Calculate the energy transferred during the test run. As you know, it is:

$F \times s$ joules.

5 Calculate the energy transferred in one second – the power. Divide the energy by the time taken. That is:

$F \times s \div t$ watts.

(If you have a calculator you would enter it like this:

$\boxed{F}$ $\boxed{\times}$ $\boxed{s}$ $\boxed{=}$ $\boxed{}$ $\boxed{\div}$ $\boxed{t}$ $\boxed{=}$ watts.)

IN YOUR NOTEBOOK

✳ Look again at your *Measuring my work* table.

✳ In the *Work* column, each number shows the amount of energy you transferred in 20 seconds. Divide each of them by 20 and you will have the amount of energy you transferred in 1 second doing that activity. These are the *powers* you developed in the different activities. Do the calculations and put the results in the right hand column.

✳ In which activity did you develop the most power? Write it down.

You can measure your power by running upstairs. Do it like this . . .

1 Find the force that you will apply as you run upstairs. This is the force needed to overcome your weight so you must weigh yourself on a set of newton bathroom scales.
2 Measure the total height of the stairs. (Hint: You do not need a very long tape-measure if the steps are all the same height, and if you know what that height is.)
3 Get someone to time you as you run upstairs.

IN YOUR NOTEBOOK

✳ Write a heading: *Measuring power*

✳ Write this statement:

Power is the energy transferred in a second. To calculate it you divide the energy transferred by the time taken. Power is measured in watts (W).

* Write a few words about measuring your own power by running upstairs. Imagine that you are explaining what happened to someone who was not there.

* Write out your results like this:

My weight = Force applied (*F*) = _______ N
Height of stairs = Distance (*s*) = _______ m
Time taken (*t*)   = _______ s
Power = $F \times s \div t$ = _______ W

* Answer these questions:

1 Is the heaviest person in your class the most powerful or the least powerful?
2 If your teacher weighs twice as much as you and has the same power, how long will it take him or her to run up the stairs?

7.17 Another look at machines

Here are some ideas for investigating another machine – one of the many types of car jack on the market. The one in the diagram is a 'hydraulic' type. The top plate of the jack rises a very small amount each time the lever is pushed down.

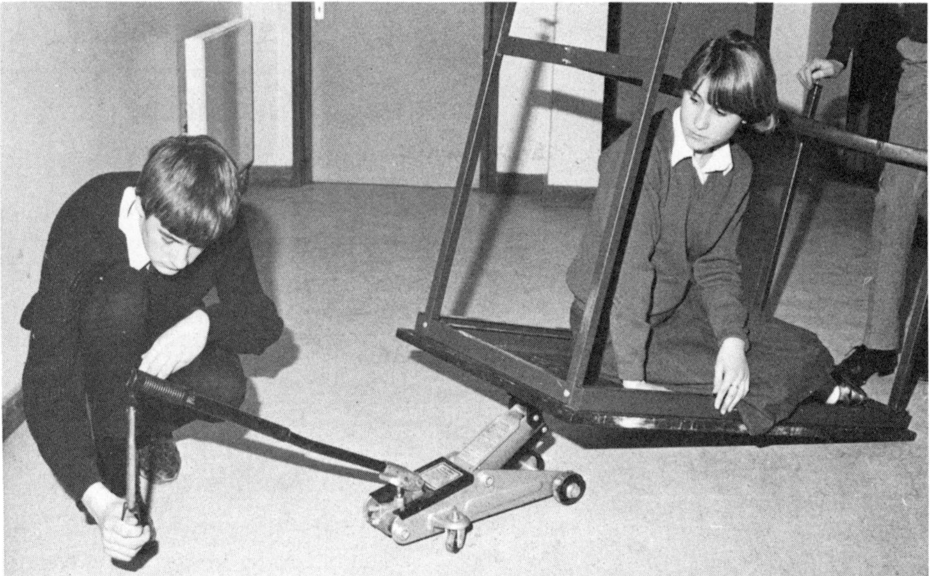

Fig. 7.35 Where would you put the scales to measure the total force pressing down on the top of the jack?

Measuring the effort force is quite easy. You use a forcemeter to pull down the lever and get someone to read it as you do so.

Measuring the load is not so easy – but you can do it with a pair of bathroom scales. Can you work out where to put the scales so they measure the total force pressing down on the top of the jack? When you think you have an answer, talk about it with your teacher.

Measure the distance you have to move the end of the lever to make the load rise a short distance – 10 cm (0.1 m) say. Do not forget to count the number of movements of the lever so that you can work out the *total* distance you move the effort.

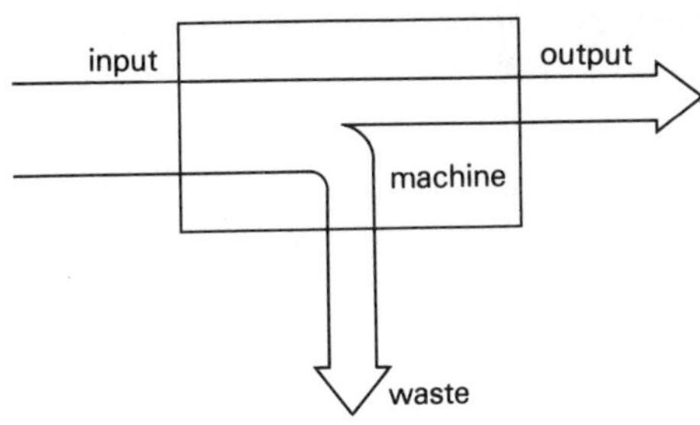

Fig. 7.36 Machines waste energy

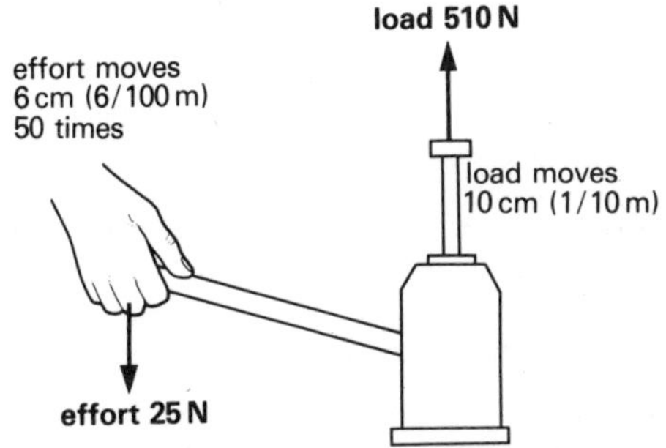

Fig. 7.37 How much energy does this machine waste?

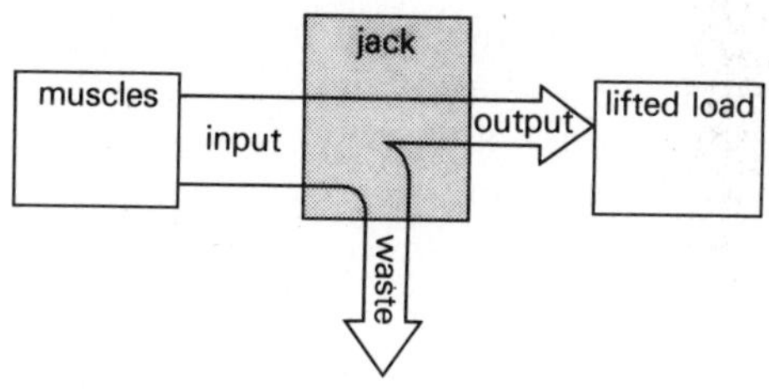

Fig. 7.38

7.18 Energy wasters – or 'you always get less out than you put in'!

When you use a force multiplying machine you have to move your effort force much further than you would have to if you used your bare hands. For example, you would not have to lift a car very far to allow a friend to change the wheel – but you would need to be very strong. Using a jack, you need much less strength, but you have to move the effort force several metres and it takes a lot longer to lift the car.

There is another disadvantage to using a machine. It has to do with energy. Like all other energy transfer devices, *machines waste energy*. Friction causes the moving parts of the machine to warm up. The air around them warms up and energy is carried away – shared out amongst the millions of molecules of the air.

How much energy do machines waste? What fraction of the input energy is lost?

You can work it out. For example, Fig. 7.37 shows some typical figures for a hydraulic jack.

You will remember from page 145 that:

energy transferred = force applied × distance moved

So you can work out the *energy input* to the machine by multiplying the effort force by the distance the effort moves. And you can work out the *energy output* by multiplying the load by the distance the load moves. So you can work out the energy wasted by the machine.

IN YOUR NOTEBOOK

✳ Write a heading: *Energy wasted by a machine*

✳ Copy Fig. 7.38. Use the figures for the hydraulic jack to calculate the input energy, output energy and wasted energy.

✳ Write these statements and fill in the gaps:

When scientists investigate a machine they work out a number which shows how good the machine is as an energy transfer device. This number is called the efficiency of the machine.

$$\text{Efficiency} = \frac{\text{Useful energy output}}{\text{Total energy input}}$$

For the hydraulic jack,

$$\text{Efficiency} = \underline{\hspace{3em}} = \qquad †$$

✳ Go back to the results of your investigations for the midget hoist (see page 138). Draw an input/output diagram for the hoist. Work out the energy amounts and fill them in.

✳ Calculate the efficiency of the hoist. Is it more efficient or less efficient than the hydraulic jack?

---

† Efficiency is usually written down as a percentage. That means as a fraction with 100 on the bottom. For example, an efficiency of 0.50 is written as:
$\frac{50}{100}$ or 50%.

7.19 Energy on the move

A steam engine is only about 7 per cent efficient (efficiency = 7/100 or 0.07). This means that out of every 100 joules of energy supplied to it as fuel, only 7 joules are transferred to its moving parts to do a useful job. 93 joules are wasted.

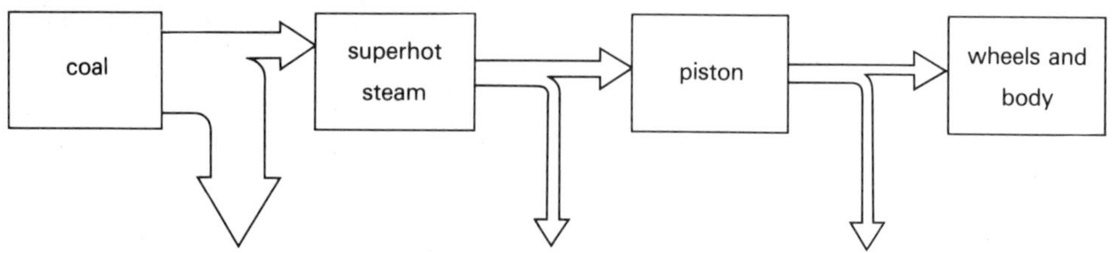

Fig. 7.39 Energy transfer diagram for a steam engine

What happens to the wasted energy? Where does it go? What does it belong to at the end?

The wasted energy is lost to the surroundings. As the fuel burns, waste gases are produced and these escape, carrying energy with them. Then friction between the moving parts of the engine makes the parts hotter. They warm up the air next to the engine. The result is that the molecules of the air carry energy away.

The energy carried away is shared with the rest of the atmosphere of Earth. We cannot get it back. It is so shared out that we cannot, with any certainty, say where it is.

Scientists have a special name for *energy on the move, energy which is being transferred because of temperature differences*, energy which is being shared out so that we cannot say what has it or where it is. That name is *heat*.

There seems to be a 'Law of Nature' about wasting energy. It is a giant pattern which says something about every energy transfer device from machines and engines to living things. It is that you *always* get less useful energy out of a machine than you put in. Every time you transfer energy some of it is always *wasted as heat*.

But not everybody has realised this . . . There are always optimists!

Since the very earliest days of engines and machines, people have tried to design ones which, once started, would go on running for ever. Such devices are called perpetual motion machines. The one in the picture was designed by an English doctor, Robert Fludd, in the year 1618. It was never built. If it had been, it could not have worked! To understand why, you need to look carefully at the way it is supposed to work. Start with the reservoir (marked A) and follow this description of how it was supposed to work:

1 Water runs out of reservoir.
2 Outflow drives waterwheel C.
3 Worm drives D, gears E, F, and G, turn shaft H and fly-wheel K.
4 Gears I and L drive shaft M.
5 Gears R and H turn device Q which is called an Archimedean screw.
6 Archimedean screw lifts water from pool near waterwheel to reservoir.
  Return to 1.

The energy transfer diagram in Fig. 7.41 sums it all up.

You have read that energy transfer devices have to lose energy as heat, so you should be able to say why the engine could not work. Think about it for a while and make a few notes. When you think you know why it wouldn't work, read the answer on page 161.

Fig. 7.42 shows another optimistic perpetual motion machine. The idea is

Fig. 7.40 Perpetual motion mill

Fig. 7.41 Energy transfer diagram for perpetual
motion mill

reservoir → outflow → waterwheel → gears

screw

gears ← driveshaft ← gears ← driveshaft and flywheel

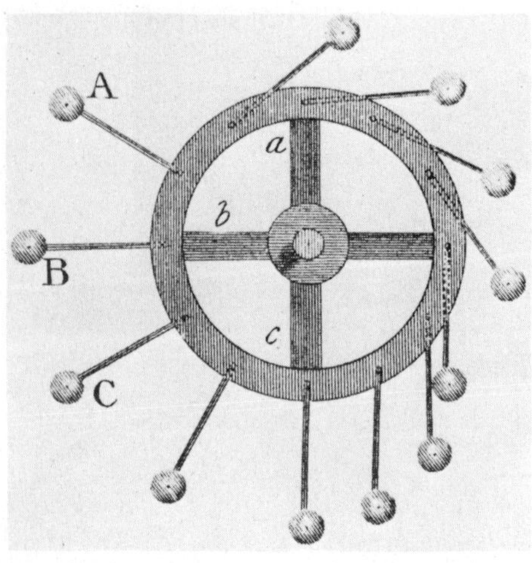

Fig. 7.42 Another perpetual motion machine

that as the weights A, B and C on the left hand side swing out, the fact that they are further from the axle will make them pull the wheel round. They will lift the weights on the right hand side until *they* reach the top and flip over – and pull the wheel round.

It doesn't work! Why not? Can you guess? (Hint: Count the weights on the left hand side of the wheel and on the right hand side.) When you think you know, talk about it with your teacher.

There have been thousands of designs for perpetual motion machines over the years. The French Patents Office refused to accept any more as long ago as 1775. The British Patents Office carried on accepting them and they are still coming in – though not as fast as they used to. For example, between 1850 and 1900 there was about one new one every month!

7.20 Increasing untidiness

Suppose you half fill a test-tube with coloured sweets so that all the yellow ones are on the bottom and all the red ones are on the top, and then shake the tube. The sweets will quickly become mixed. It does not matter how long you shake the tube, the sweets will never unmix themselves into two layers.

If you leave an ice cube on a saucer in a warm place, the cube will melt. If you put the saucer in the fridge the water will freeze again but it will never rearrange itself into a neat cube.

If you break an egg and leave it, shell and all, in a bowl, you will never come back to find that it has joined itself up again to make a complete egg.

You have known about things like this since you were very small. You were not very old before you knew that iron rusts and rust does not turn back into iron again; and that springs unwind and never wind themselves up; and that people get older and never younger.

All of these things are connected by an important giant pattern. The pattern includes the idea that machines and engines must waste energy. In simple words the pattern says that our universe is getting more and more untidy and mixed up as time goes by. Scientists call the pattern *The Second Law of Thermodynamics or The Entropy Law*.

Scientists measure the untidiness of a collection of things – they say the *degree of disorder* or the *entropy* of the collection – by asking the question:

How easy is it to say where everything is in this collection?

If everything is in its 'right' place and if it is easy to say where everything is, then it is a 'tidy' collection (scientists would say a '*low entropy system*').

The Entropy Law says that *whenever things change, the entropy of the universe increases*.

Before the fire in Fig. 7.43 was lit we could say just where the energy was. We could point to a piece of coal and say: 'That has so much energy – it has so much ability to cause movement'. When the fire has gone out, the energy will have been spread out all over the place and we will not be able to get it back. The chemical substances in the coal will have been spread out too as smoke, soot, and waste gases. We will not be able to get them back. By lighting a fire to transfer energy from coal to our house we have made the whole universe less tidy! It is the same whenever we arrange an energy transfer.

Living things *seem* to be able to put the Law of Entropy into reverse. Trees can transfer energy from sunlight to the chemical substances they make in their leaves. They can collect up scattered chemicals from soil, water, and the air and store them tidily in their tissues.

Fig. 7.43 A coal fire burning in a home

Fig. 7.44 Giant redwood trees. A massive reversal of the Law of Entropy?

Humans can think of clever ways to tidy things up – feeding scrambled eggs to chickens to turn them back into eggs perhaps? Or inventing and making fridges to turn water into neat little cubes of ice.

But trees and humans cannot win. To do any of these things we must always waste so much energy that the tidiness we produce is easily outweighed by the untidiness produced in the rest of the universe.

Checkout

Keywords

efficiency	heat
energy	inertia
energy transfer diagram	machine
engine	mass
entropy	power
force	speed
friction	weight
gravity	

Patterns

1 Things which can move tend to stay still and it takes a force to start them moving. Once they are moving they tend to go on in a straight line at a steady speed. (This property is called *inertia*. The numerical measure of inertia is *mass*.)

2 Friction forces depend on the nature of the surfaces in contact, and on the forces pressing the surfaces together.

3 Everything in the universe attracts everything else with a force which depends on its mass.

4 Machines *either* give an output force which is greater than the input force, *or* give an output distance moved which is greater than the input distance moved. No machine can do both at the same time.

5 Energy transferred to moving things or to things which will move if released can be measured using the formula:

Force applied × distance moved.

(The number which is obtained is called the *work* measured in joules.)

6 In comparing the usefulness of energy transfer devices, the *rate* of energy transfer (*power*), is usually more relevant than the total energy transferred.

7 All machines and engines waste energy. That is, whenever energy is transferred, only part of the input produces the desired change. So:

$$\text{efficiency} = \frac{\text{useful energy output}}{\text{total energy input}}$$

is always less than 1.

8 Whenever things change, the degree of disorder – the *entropy* – of the universe increases.

True or false?

1 You could jump much higher on the Moon than you can on Earth because your mass would be only one sixth of your Earth mass.

2 There are no mistakes in the following table:

Quantity	Unit
Mass	kg
Force	N
Energy	J
Power	W

3 An artificial satellite moving in a circular orbit at a steady speed has no forces acting on it.

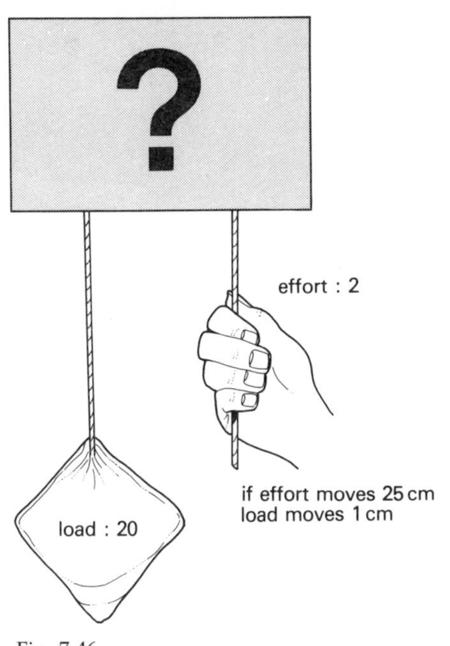

Fig. 7.46

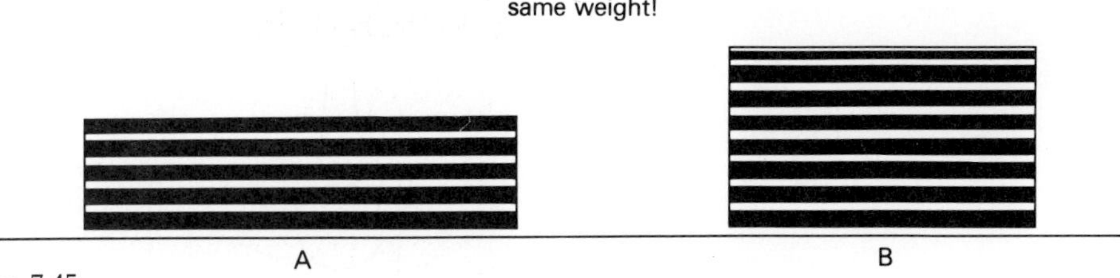

Fig. 7.45

4 The more petrol a car has in its tank, the greater the force the wheels must apply to make it turn a corner.

5 Puck A in Fig. 7.45 would be much harder to start moving than puck B.

6 Friction forces help the machine called a screw jack to do its job properly.

7 If Jim transfers 800 J of energy from his food in lifting a 120 N parcel on to a shelf 2 metres high, he will have done 240 J of work.

8 Susan weighs half as much as her father and can run upstairs in half the time it takes him, so her power output is four times as great as his.

9 Looked at as a machine, a fishing rod is a force multiplying device.

10 The efficiency of the mystery machine in Fig. 7.46 is 0.4 (40%).

Problems

1 *Read this passage and try to answer the questions.*

In a cold remote part of the world, a small town is to be built. It will house the construction workers for a new nuclear power station. Natural gas can be obtained from local wells but the town has no electricity supplies. Two alternative systems for heating the houses have been suggested and a choice has to be made. The systems are shown in Figs. 7.47 and 7.48.

System A

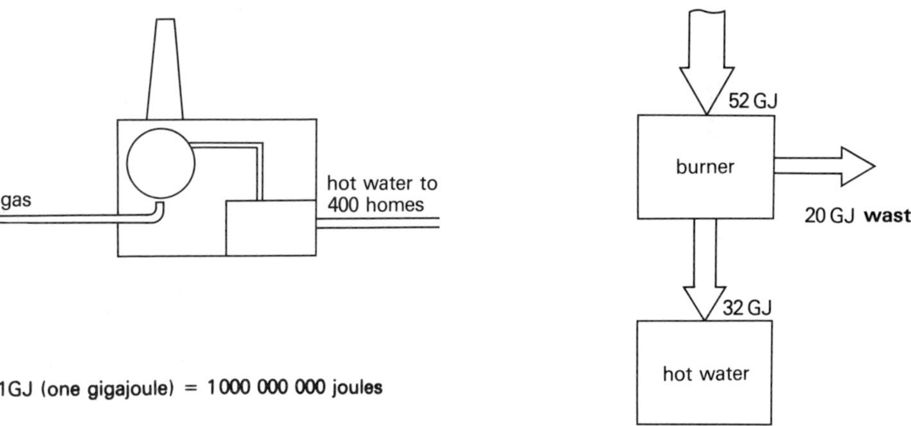

Fig. 7.47

SYSTEM A

Natural gas is burned in a central power station and hot water is piped to all the homes.

SYSTEM B

Natural gas is burned to produce steam in a central power station. The steam drives a turbo-generator. Electricity from the generator is supplied to immersion heaters in the household central heating tanks.

(a) Which system would you recommend? (Give three reasons.)

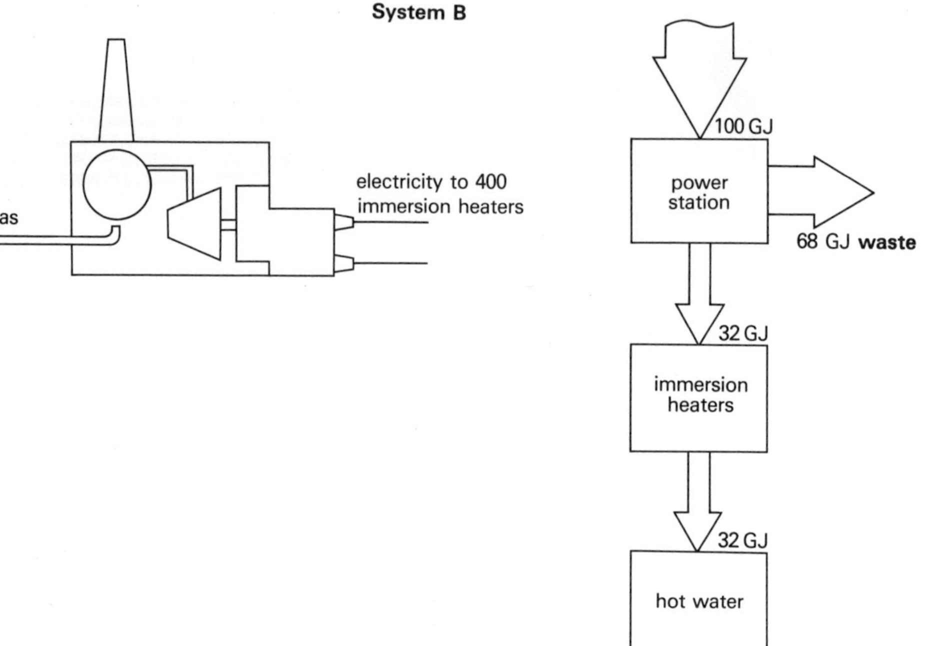

Fig. 7.48

(b) Which system is the most efficient according to the energy transfer
    diagrams? (Hint: energy input = energy transferred from the gas; useful
    energy output = energy transferred to the water.)
(c) Both systems waste energy in ways which are not shown in the
    diagrams. What are they?
(d) In which system is this 'extra loss' likely to be greatest? Which system
    would cool down most quickly if the gas supply was cut off? (Give your
    reasons.)
(e) Can you think of a system which would waste less energy than either A
    or B? Draw a diagram of your system.
(f) If System B was built, it could supply the people in the town with more
    than hot water. What else would it supply?
(g) Which system would be the best to have when the nuclear power station
    was finished?

2 *On the 8th July 1981 a solar-powered aircraft designed by the American engineer
Dr Paul MacCready became the first to cross the English Channel. A fortnight
earlier New Scientist magazine had published an article about the 'plane. The
passage below is taken from that article.*
Read the passage and try to answer the questions.
. . . The solar cells that produce electricity are so small that you must have
thousands of them. Solar Challenger has over 16 000 and they have cost
MacCready $35 000 just to wire up and test. (About £40 000 worth were
loaned to MacCready by NASA for the flight). . . . the cells cover most of
the wings and tail and produce a maximum of 1.1 kW per square metre.
. . . MacCready admits that in practical terms, they do not compare with
the single piston engine used on modern light aircraft (which typically) . . .
gives about 112 kW of power . . . . The maximum power you could get out of
the same plane by covering its wings with present day solar cells is a
meagre 1.1 kW. . . . Solar Challenger has four times more power . . . . The
pitch of the propeller is altered by the pilot to match the designed power (to
that) available from the cells. So in hazy conditions when less than the full
4.5 kW is available the propeller's pitch is not so great that it stops or stalls.
The pilot watches a watt-meter to tell him how to select the optimum pitch.

Fig. 7.49 Solar Challenger

(a) What would be the advantages of solar-powered aircraft (over piston-engined types)? Try to think of three.
(b) What are the present disadvantages? Try to think of two.
(c) What would be the most serious problem if you wanted to fly a solar-powered aircraft regularly in the UK? Can you think of a solution to the problem?
(d) The Solar Challenger's wings and tail are shorter and stubbier and flatter on the top than those of an ordinary aircraft. Why do you think that is?
(e) Can you think of a kind of aircraft which would be particularly suited to solar power? (Hint: it is very easy to get such an aircraft off the ground!)
(f) Suppose the efficiency of the solar cells was 15%. How much energy would they collect from sunlight every second? (What would be their energy input?)
(g) With one of the pilots, Janice Brown, aboard, the Solar Challenger weighs about 1350 N. How much energy would be needed to get it 3 metres into the air?

Answers

Below are the answers to the questions on the pages shown.

PAGE 4

1 Double loop **2** Arch **3** Whorl
4 Whorl

PAGE 8

Cream, ginger biscuit

PAGE 9

Key:
A7, B4, C6, D5, E3, F2, G8, H1, I9

PAGE 17

The carbonates fizzed when you added the acids. That was because a gas was being 'given off'. So the diagram should look like this.

$$\boxed{\text{ACID}} + \boxed{\text{CARBONATE}} \rightarrow \boxed{\text{GAS}}$$

PAGE 24

1 It was a bright sunny day in Ascot with just a few fluffy clouds in the sky.
2 'Rain in Cornwall' would be a good guess – it was right under the front. But – in fact – the cloud was so low there that the county was covered in a blanket of wet fog!

PAGE 36

The magma which is thrown out of the crater will cool suddenly. No crystals will form in it – it will be glassy. Rock which gets trapped in one of the 'branches' (called dykes) of the volcano will cool more slowly.

PAGE 37

The oldest rocks are the lowest rocks. Just like a pile of letters in a letter box, the ones which were laid down first are at the bottom.

Sometimes, folding and faulting can turn the layers over and mix them up, so that geologists find it difficult to decide which is the oldest. If they are in doubt they look at the layers separately. The biggest bits of rock in a layer can show the parts which were on the bottom when the rock formed. (Big bits of rock settle down more quickly than small bits.)

PAGE 55

Falcon – (b)
Humming bird – (c)
Sparrow – (d)
Flamingo – (a)

PAGE 56

Eagle – (d)
Sparrow – (a)
Duck – (b)
Woodpecker – (c)

PAGE 128

Puck will hit dots 8 and 9.

PAGE 145

3 parcels would be put on shelf 4 to do just the same amount of work.

PAGE 146

1 25 J.   **2** 150 J.   **3** 125 J.
4 50 J each by putting them on shelf 2.
5 One parcel on shelf 3 or 3 parcels on shelf 1.
6 Both people transferred the same amount of energy to *parcels*, but the first person had to move herself very much further. She had to transfer energy from her food through her muscles to her whole body as she climbed the ladder.

PAGE 152

Suppose you started things going by filling the reservoir with water. By doing so you would put a certain amount of energy into the system and it would start working (at least, it might!). However, *every arrow in the transfer diagram should have a side arrow leading out of it to show energy wasted.* As soon as the wastage amounted to the energy you put in at the beginning the system would stop.

Index